Introduction to Mycology

Introduction to Mycology

Penelope Goodman

STATES
ACADEMIC PRESS
www.statesacademicpress.com

States Academic Press,
109 South 5th Street,
Brooklyn, NY 11249, USA

Visit us on the World Wide Web at:
www.statesacademicpress.com

ISBN: 978-1-63989-307-2 (Hardback)

Cataloging-in-Publication Data

Introduction to mycology / Penelope Goodman.
p. cm.
Includes bibliographical references and index.
ISBN 978-1-63989-307-2
1. Mycology. 2. Fungi. 3. Microbiology. 4. Botany. I. Goodman, Penelope.
QK603 .I58 2022
579.5--dc23

TABLE OF CONTENTS

PREFACE

The purpose of this book is to help students understand the fundamental concepts of this discipline. It is designed to motivate students to learn and prosper. I am grateful for the support of my colleagues. I would also like to acknowledge the encouragement of my family.

Fungus is any member of eukaryotic organisms which include yeasts, mushrooms, molds, mildews, rusts and smuts. Mycology is the field of biology which focuses on the study of fungi. It includes fungi's biochemical and genetic properties, taxonomy, and diseases. Mycotoxicology focuses on the analysis and study of fungal toxins called mycotoxins. Ethnomycology is an interdisciplinary field concerned with the study of sociological impact and historical uses of fungi. Fungi are used by forensic experts in criminal examinations. This is known as forensic mycology. Medicinal mycology is concerned with study of pathogenic fungi, fungal diseases along with their ecology and epidemiology. Basidiobolomycosis, valley fever, cryptococcosis and fungal meningitis are some of the common fungal diseases. This book is a compilation of chapters that discuss the most vital concepts in the field of mycology. It presents this complex subject in the most comprehensible and easy to understand language. Those in search of information to further their knowledge will be greatly assisted by this textbook.

A foreword for all the chapters is provided below:

Chapter - What is Mycology?

A fungus is a member of the group of eukaryotic organisms including microorganisms like yeasts, molds, mushrooms, etc. The branch of biology that deals with the study of fungi is termed as mycology. This is an introductory chapter which will briefly introduce about mycology.

Chapter - Branches of Mycology

Mycology can be divided into a number of branches to focus on specific areas of the field. Some of them are lichenology, mycotoxicology, ethnomycology, forensic mycology, etc. This chapter closely examines the varied branches of mycology to provide an extensive understanding of the subject.

Chapter - Biological Classification of Fungi

Fungi can also be classified on the basis of their biological characteristics. Biological classification of fungi includes basidiomycota, ascomycota, neocallimastigomycota, blastocladiomycota, glomeromycota, chytridiomycota, microsporidia, etc. This chapter has been carefully written to provide an easy understanding of such biological classifications of fungi.

Chapter - Types of Fungi

There are different types of fungi in the world which includes mushroom, rust, puffball, polypore, morchella, truffle, yeast, mildews, molds, etc. The topics elaborated in this chapter will help in gaining a better perspective of the aspects related to these different types of fungi.

Chapter - Reproduction and Spores of Fungi

Fungal reproduction can take place asexually through fragmentation, budding, and producing spores. Some of the fungal spores are sporangiospore, zygospore, ascospore, basidiospore, aeciospore, ascus, oospore, etc. This chapter sheds light on the subject of fungal reproduction and spores for a thorough understanding of it.

Chapter - Fungal Diseases

Fungal diseases occur when a harmful fungus enters the immune system of the body. Dermatophytosis, Fungal keratitis, sporotrichosis, blastomycosis, histoplasmosis, aspergillosis, coccidioidomycosis, paracoccidioidomycosis, etc. are a few of the fungal diseases. This chapter discusses the symptoms, sources, diagnosis and treatment of various fungal diseases in detail.

Penelope Goodman

What is Mycology?

CHAPTER 1

A fungus is a member of the group of eukaryotic organisms including microorganisms like yeasts, molds, mushrooms, etc. The branch of biology that deals with the study of fungi is termed as mycology. This is an introductory chapter which will briefly introduce about mycology.

FUNGI

Fungi are eukaryotic microorganisms. Fungi can occur as yeasts, molds, or as a combination of both forms. Some fungi are capable of causing superficial, cutaneous, subcutaneous, systemic or allergic diseases. Yeasts are microscopic fungi consisting of solitary cells that reproduce by budding. Molds, in contrast, occur in long filaments known as hyphae, which grow by apical extension. Hyphae can be sparsely septate to regularly septate and possess a variable number of nuclei. Regardless of their shape or size, fungi are all heterotrophic and digest their food externally by releasing hydrolytic enzymes into their immediate surroundings (absorptive nutrition). Other characteristics of fungi are the ability to synthesize lysine by the L-α-adipic acid biosynthetic pathway and possession of a chitinous cell wall, plasma membranes containing the sterol ergosterol, 80S rRNA, and microtubules composed of tubulin.

Physiology

Fungi can use a number of different carbon sources to meet their carbon needs for the synthesis of carbohydrates, lipids, nucleic acids, and proteins. Oxidation of sugars, alcohols, proteins, lipids, and polysaccharides provides them with a source of energy. Differences in their ability to utilize different carbon sources, such as simple sugars, sugar acids, and sugar alcohols, are used, along with morphology, to differentiate the various yeasts. Fungi require a source of nitrogen for synthesis of amino acids for proteins, purines and pyrimidines for nucleic acids, glucosamine for chitin, and various vitamins. Depending on the fungus, nitrogen may be obtained in the form of nitrate, nitrite, ammonium, or organic nitrogen; no fungus can fix nitrogen. Most fungi use nitrate, which is reduced first to nitrite (with the aid of nitrate reductase) and then to ammonia.

Nonfungal organisms, including bacteria, synthesize the amino acid lysine by the meso-α, ε-diaminopimelic acid pathway (DAP pathway), whereas fungi synthesize lysine by only the L-α-adipic acid pathway (AAA pathway). Use of the DAP pathway is one of the reasons microorganisms previously considered to be fungi, such as the myxomycetes, oomycetes, and hyphochytrids, are no longer classified as fungi. The DAP and AAA biosynthetic pathways for lysine synthesis represent dichotomous evolution.

Structure

Cell Wall

The rigid cell wall of fungi is a stratified structure consisting of chitinous microfibrils embedded in a matrix of small polysaccharides, proteins, lipids, inorganic salts, and pigments that provides skeletal support and shape to the enclosed protoplast. Chitin is a (β1–4)-linked polymer of N-acetyl-D-glucosamine (GlcNAc). It is produced in the cytosol by the transfer of GlcNAc from uridine diphosphate GlcNAc into chains of chitin by chitin synthetase, which is located in the cytosol in organelles called chitosomes. The chitin microfibrils are transported to the plasmalemma and subsequently integrated into the new cell wall.

The major polysaccharides of the cell wall matrix consist of noncellulosic glucans such as glycogen-like compounds, mannans (polymers of mannose), chitosan (polymers of glucosamine), and galactans (polymers of galactose). Small amounts of fucose, rhamnose, xylose, and uronic acids may be present. Glucan refers to a large group of D-glucose polymers having glycosidic bonds. Of these, the most common glucans composing the cell wall have the β-configuration. Polymers with (β1–3)- and (β1–6)-linked glucosyl units with various proportions of 1–3 and 1–6 linkages are common. Insoluble β-glucans are apparently amorphous in the cell wall. In Paracoccidioides brasiliensis, the hyphal cell wall consists of a single, 80- to 150-nm layer composed of chitin and β-glucan. In contrast, the 200- to 600-nm-thick yeast cell wall has three layers. The inner surface is chitinous, containing some α-glucan, and the outer layer contains α-glucan. It has been suggested that the (α1–3)-glucan occurs in a microfibrillar form in P brasiliensis and Histoplasma capsulatum.

Many fungi, especially the yeasts, have soluble peptidomannans as a component of their outer cell wall in a matrix of α- and β-glucans. Mannans, galactomannans, and, less frequently, rhamnomannans are responsible for the immunologic response to the medically important yeasts and molds. Mannans are polymers of mannose or heteroglucans with α-D-mannan backbones. Structurally, mannan consists of an inner core, outer chain, and base-labile oligomannosides. The outer-chain region determines its antigenic specificity. Determination of mannan concentrations in serum from patients with disseminated candidiasis has proven a useful diagnostic technique.

Cryptococcus neoformans produces a capsular polysaccharide composed of at least three distinct polymers: glucuronoxylomannan, galactoxylomannan, and mannoprotein. On the basis of the proportion of xylose and glucuronic acid residues, the degree to which mannose has side-chain substituents, and the percentage of O-acetyl attachments of the capsular polysaccharides, isolates of C neoformans can be separated into four antigenic groups designated A, B, C, and D. The capsule is antiphagocytic, serves as a virulence factor, persists in body fluids, and allows the yeast to avoid detection by the host immune system.

In addition to chitin, glucan, and mannan, cell walls may contain lipid, protein, chitosan, acid phosphatase, α-amylase, protease, melanin, and inorganic ions such as phosphorus, calcium, and magnesium. The outer cell wall of dermatophytes contains glycopeptides that may evoke both immediate and delayed cutaneous hypersensitivity. In the yeast Candida albicans, for example, the cell wall contains approximately 30 to 60 percent glucan, 25 to 50 percent mannan (mannoprotein), 1 to 2 percent chitin (located primarily at the bud scars in the parent yeast cell wall), 2 to 14 percent lipid, and 5 to 15 percent protein. The proportions of these components vary greatly from fungus to fungus. Summarizes the relationship between cell wall composition and taxonomic grouping of the fungi.

Principal cell wall polymer	Taxonomic Group	Examples
Chitin-chitosan	Zygomycetes	*Rhizopus arrhizus*
Chitin-glucan	Ascomycetes (mycelial)	*Pseudallescheria boydii*
	Basidiomycetes (mycelial)	*Schizophyllum commune*
Glucanmannan	Ascomycetes (yeast)	*Sacchamomyces cerivisea*
Chitin-mannan	Fungi imperfecti	*Candida albicans*
	Basidiomycetes (yeast)	*Filobasidiella neoformans*

Cell Wall Composition and Taxonomic Classification of Representative Medically Important Fungi.

Plasma Membrane

Fungal plasma membranes are similar to mammalian plasma membranes, differing in having the nonpolar sterol ergosterol, rather than cholesterol, as the principal sterol. The plasma membrane regulates the passage of materials into and out of the cell by being selectively permeable. Membrane sterols provide structure, modulation of membrane fluidity, and possibly control of some physiologic events.

The plasma membrane contains primarily lipids and protein, along with small quantities of carbohydrates. The major lipids are the amphipathic phospholipids and sphingolipids that form the lipid bilayer. The hydrophilic heads are toward the surface, and the hydrophobic tails are buried in the interior of the membrane. Proteins are interspersed in the bilayer, with peripheral proteins being weakly bound to the membrane. In contrast, integral proteins are tightly bound. The lipoprotein structure of the membrane provides an effective barrier to many types of molecules. Molecules cross the membrane by either diffusion or active transport. The site of interaction for most antifungal agents is the ergosterol in the membrane or its biosynthetic pathway. Polyene antifungal agents such as amphotericin B bind to ergosterol to form complexes that permit the rapid leakage of the cellular potassium, other ions, and small molecules. The loss of potassium results in the inhibition of glycolysis and respiration.

Several antifungal agents interfere with ergosterol synthesis. The first step in the synthesis of both ergosterol and cholesterol is demethylation of lanosterol. The necessary enzymes are associated with fungal microsomes, which contain an electron transport system analogous to the one in liver microsomes. Cytochrome P_{450} catalyzes the 1 4-α-demethylation of lanosterol, an essential step in the synthesis of ergosterol. The imidazole and triazole antifungal agents interfere with cytochrome P_{450}-dependent 1 4-α-demethylase, which inhibits the formation of ergosterol. This results in plasma membrane permeability changes and inhibition of growth. Ergosterol may also be involved in regulating chitin synthesis. Inhibition of ergosterol synthesis by antifungal agents can result in a general activation of chitin synthetase zymogen, leading to excessive chitin production and abnormal growth.

Microtubules

Fungi possess microtubules composed of the protein tubulin. This protein consists of a dimer composed of two protein subunits. Microtubules are long, hollow cylinders approximately 25 nm in diameter that occurs in the cytoplasm as a component of larger structures. These structures are involved in the movement of organelles, chromosomes, nuclei, and Golgi vesicles containing cell wall precursors.

Microtubules are the principal components of the spindle fibers, which assist in the movement of chromosomes during mitosis and meiosis. When cells are exposed to antimicrotubule agents, the movement of nuclei, mitochondria, vacuoles, and apical vesicles is disrupted. Griseofulvin, which is used to treat dermatophyte infections, binds with microtubule-associated proteins involved in the assembly of the tubulin dimers. By interfering with tubulin polymerization, griseofulvin stops mitosis at metaphase. The destruction of cytoplasmic microtubules interferes with the transport of secretory materials to the cell periphery, which may inhibit cell wall synthesis.

The fungal nucleus is bounded by a double nuclear envelope and contains chromatin and a nucleolus. Fungal nuclei are variable in size, shape, and number. The DNA and associated proteins occur as long filaments of chromatin, which condenses during nuclear division. The number of chromosomes varies with the particular fungus. Within the cell, 80 to 99 percent of the genetic material occurs in chromosomes as chromatin, and approximately 1 to 20 percent in the mitochondria. In some isolates of Saccharomyces cerevisiae, up to 5 percent of their DNA can be found in nuclear plasmids. When the DNA helix unwinds, one strand serves as the template for the synthesis of rRNA, tRNA, and mRNA. mRNA passes into the cytoplasm and attaches to one of the ribosomes, which are complexes of RNA and protein that serve as sites for the synthesis of protein.

Morphology

Yeasts

Yeasts are fungi that grow as solitary cells that reproduce by budding. Yeast taxa are distinguished on the basis of the presence or absence of capsules, the size and shape of the yeast cells, the mechanism of daughter cell formation (conidiogenesis), the formation of pseudohyphae and true hyphae, and the presence of sexual spores, in conjunction with physiologic data. Morphology is used primarily to distinguish yeasts at the genus level, whereas the ability to assimilate and ferment various carbon sources and to utilize nitrate as a source of nitrogen are used in conjunction with morphology to identify species.

Yeasts such as C albicans and Cryptococcus neoformans produce budded cells known as blastoconidia. The formation of blastoconidia involves three basic steps: bud emergence, bud growth, and conidium separation. During bud emergence, the outer cell wall of the parent cell thins. Concurrently, new inner cell wall material and plasma membrane are synthesized at the site where new growth is occurring. New cell wall material is formed locally by activation of the polysaccharide synthetase zymogen. The process of bud emergence is regulated by the synthesis of these cellular components as well as by the turgor pressure in the parent cell. Mitosis occurs, as the bud grows, and both the developing conidium and the parent cell will contain a single nucleus. A ring of chitin forms between the developing blastoconidium and its parent yeast cell. This ring grows in to form a septum. Separation of the two cells leaves a bud scar on the parent cell wall. The bud scar contains much more chitin than does the rest of the parent cell wall. When the production of blastoconidia continues without separation of the conidia from each other, a pseudohypha, consisting of a filament of attached blastoconidia, is formed. In addition to budding yeast cells and pseudohyphae, yeasts such as C albicans may form true hyphae.

Candida Albicans

Candida albicans may form a budding yeast, pseudohyphae, germ tubes, true hyphae, and chlamydospores. A number of investigators are interested in germ tube formation because it represents a transition between a yeast and a mold. Generally, either low temperature or pH favors the development of a budding yeast. Other substances such as biotin, cysteine, serum transferrin, and zinc stimulate dimorphism in this yeast.

Approximately 20 percent of the C albicans yeast cell wall is mannan, whereas the mycelial cell wall contains a substantially smaller amount of this sugar. Candida albicans has three serotypes, designated A, B, and C. These are distinguished from each other on the basis of their mannans. The antigenic determinant for serotype A is its mannoheptaose side chain.

In serotype B, it is the mannohexaose side chain. Serotype B tends to be more resistant to 5-fluorocytosine than is serotype A. Glucans with (β1–3) and (β1–6)-linked groups compose about 50 to 70 percent of the yeast cell wall. It has been suggested that these glucans may impede the access of amphotericin B to the plasma membrane.

Molds

Molds are characterized by the development of hyphae, which result in the colony characteristics seen in the laboratory. Hyphae elongate by a process known as apical elongation, which requires a careful balance between cell wall lysis and new cell wall synthesis. Because molds are often differentiated on the basis of conidiogenesis, structures such as conidiophores and conidiogenous cells must be carefully evaluated. Some molds produce special sac-like cells called sporangia, the entire protoplasm of which becomes cleaved into spores called sporangiospores. Sporangia are typically formed on special hyphae called sporangiophores.

Dimorphism

A number of medically important fungi express themselves phenotypically as two different morphologic forms, which correlate with the saprophytic and parasitic modes of growth. Such fungi are called dimorphic fungi. Some researchers restrict the term to pathogens that grow as a mold at room temperature in the laboratory and as budding yeast or as spherules either in tissue or at 37 °C. In contrast, others use dimorphic for any fungus that can exist as two different phenotypes, regardless of whether it is pathogenic. We prefer to use the term "dimorphic" to describe fungi that typically grow as a mold in vitro and as either yeast cells or spherules in vivo. Examples of medically important dimorphic fungi include Blastomyces dermatitidis (hyphae and yeast cells) and Coccidioides immitis(hyphae and spherules).

ROUP	ATHOGEN	OCATION	ISEASE
1)Superficial mycosis	Piedraia hortae .	scalp	Black piedra.
	Trichosporon beigelii.	Beard ,mustache	White piedra.
	Malassezia furfur.	Trunk ,neck ,face, arms.	Tinea versicolor.
2)Cutaneous mycosis	Trichophyton mentagrophytis ,T.verrucossum, T.rubrum.	Beard hair	Tinea barbae.
	Trichophyton ,Microsporum canis.	Scalp hair	Tinea captis
	Trichophyton rubrum ,T.mentagrophytes ,Microsporum canis.	Smooth or bare parts of skin	Tinea corporis
	T.Rubrum , T.mentagrophytes Epidermophyton floccosum .	Groin, buttocks	Tinea cruris(jock itch).
	T.rubrum ,T.mentagrophytes ,E.floccosum.	feet	Tinea pedis
	T.rubrum, T. mentagophytes, E floccosum.	nails	Tinea unguium (oncomycosis).

Major Medically Important Fungi.

A number of external factors contribute to the expression of dimorphism. Increased incubation temperature is the single most important factor. Increased carbon dioxide concentration, which probably affects the oxidation-reduction potential, enhances the conversion of the mycelial form to the tissue form in C immitis and Sporothrix schenckii. pH affects the development of the yeast form in some fungi, and cysteine or other sulfhydryl-containing compounds affects it in others. Some fungi require a combination of these factors to induced dimorphism.

Blastomyces Dermatitidis

The conversion of the mycelial form of Blastomyces dermatitidis to the large, globose, thick-walled, broadly based budding yeast form requires only increased temperature. Hyphal cells enlarge and undergo a series of changes resulting in the transformation of these cells into yeast cells. The cells enlarge, separate, and then begin to reproduce by budding. The yeast cell wall contains approximately 95 percent (α1–3)-glucan and 5 percent (β1–3)-glucan. In contrast, the mycelial cell wall contains 60 percent (β1–3)-glucan and 40 percent (α1 -3)-glucan.

Coccidioides Immitis

Coccidioides immitis is a unique dimorphic fungus because it produces spherules containing endospores in tissue, and hyphae at 25°C. Increased temperature, nutrition, and increased carbon dioxide are important for the production of sporulating spherules. A uninucleate arthroconidium begins to swell and undergo mitosis to produce additional nuclei. Once mitosis stops, initiation of spherule septation occurs. The spherule is segmented into peripheral compartments with a persistent central cavity. Uninucleate endospores occurring in packets enclosed by a thin membranous layer differentiate within the compartments. As the endospores enlarge and mature, the wall of the spherule ruptures to release the endospores. Pairs of closely appressed endospores that have not completely separated from each other may resemble the budding yeast cells of B dermatitidis.

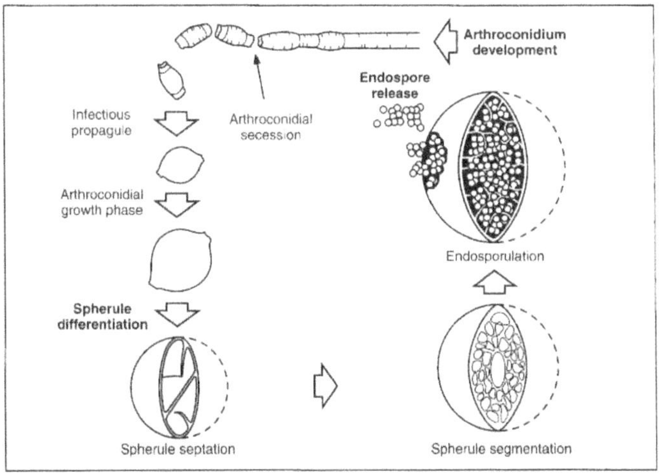

Development of the spherule of *Coccidioides immitis* from an arthroconidium.

Histoplasma Capsulatum

Dimorphism in *Histoplasma capsulatum* involves three stages. In the first stage, induced by an increase in temperature, respiration ceases and the level of cytochromes decreases. During the second stage of the mycelial-to-yeast conversion, cysteine or other sulfhydryl-containing compounds are required. Shunt pathways are initiated that restore the appropriate cytochrome levels, which supply the needed ATP. Cysteine is also required for the yeast form to grow. The final stage is characterized by normal cytochrome levels and respiration as the yeast grows and reproduces. The conversion of terminal or intercalary hyphal cells to a yeast form requires 3 to 14 days. In tissue, *H capsulatum* proliferates within giant cells.

Yeast cells of *H capsulatum* have been divided into two chemotypes. Chemotype 1, which correlates with serotypes 1, 2, and 3, contains large amounts of (β1–3)-glucans and small amounts of chitin. Chemotype 2 contains (α1–3)-glucans, a little (β1–3)-glucan, and more chitin. Chemotype 2 correlates with serotypes 1 and 4. It is difficult to assess the importance of the chemotypes, because only a few isolates have been studied.

Paracoccidioides Brasiliensis

A great deal of work has been done with the mycelial-to-yeast conversion of *Paracoccidioides brasiliensis*. In tissue the yeast is characterized by multiple budding. Series of smaller yeast daughter cells attached by narrow tubular necks are formed around a large central cell. Hyphal cells first swell and then separate from each other. The separated cells begin to bud, resulting in a yeast growth. As with *H capsulatum*, growth stops briefly as a result of increased temperature. At 37°C, synthesis of (β1–3)-glucan decreases and the hyphal cell wall softens. The hyphal cells then separate, and (α1–3)-glucans are formed as a layer on the outer cell wall surface of the yeast cells.

The yeast cell wall of *P brasiliensis* has three layers and is approximately 200 to 600 nm thick. The inner surface has chitin and some (β1–3)-glucan. The outer cell wall layer consists of (α1–3)-glucans. In contrast, the mycelial cell wall consists of one layer that is 80 to 150 nm thick, composed of chitin and (β1–3)-glucans. The (α1–3)-glucan of *P brasiliensis* is an important virulence factor. Only the yeast form has this glucan. When the α-linked glucan is absent, pathogenicity is attenuated; regeneration of α-glucan results in increased virulence.

Penicillium Marneffei

Penicillium marneffei is a dimorphic fungus that is becoming extremely important as a pathogen in AIDS patients living in Southeast Asia. The fungus has been recovered from soil associated with plants such as bamboo. In tissue, the fungus forms yeast cells that divide by fission. Like H. capsulatum, they proliferate within giant cells.

Sporothrix Schenckii

The last dimorphic fungus to be considered is *Sporothrix schenckii.* In this species, mycelial-to-yeast conversion is enhanced by increased carbon dioxide, increased temperature, and nutrition. The yeast form readily appears at 37°C and 5 percent carbon dioxide. It has been suggested that some product of carbon dioxide fixation may be required for development of the yeast form. Unlike the other dimorphic fungi capable of producing a yeast form, *S schenckii* initially produces yeast cells by direct budding from hyphae. In addition, the cell wall chemistry of the hyphal and yeast forms is similar. Glucans having (β1–3)-, (β1–4)-, and (β1–6)-linkages are present in addition to chitin. Rhamnomannan is the major antigenic determinant. The yeast cell wall is thought to contain more peptidorhamnomannan than the hyphal cell wall.

Propagules (Spores and Conidia)

Spores can be produced either asexually or sexually. Asexual spores are always formed in a sporangium following mitosis and cytoplasmic cleavage. The number of sporangiospores and their arrangement in the sporangium are used to differentiate the various zygomycetes. Sexual spores occur following meiosis. Ascospores are formed in a saclike cell (called an ascus) by free-cell formation, basidiospores form on basidia, and zygospores form within zygosporangia. Oospores are sexual spores that are produced by one group of fungi that will not be considered because they are medically unimportant. Sexual spores are rarely seen in clinical isolates because most fungi are heterothallic (i.e., sexually self-sterile). Typically, only one of the two mating types is isolated from a particular clinical specimen. When homothallic isolates are recovered in the clinical laboratory, they often produce sexual spores because they are sexually self-fertile.

Conidia are always asexual in origin and develop in any manner that does not involve cytoplasmic cleavage. The ontogeny of conidia (conidiogenesis) and their arrangement, color, and septation are used to differentiate the various genera of molds. Some fungi have melanin in the cell wall of the conidia, the hyphae, or both. Such fungi are considered to be dematiaceous. Many of the name changes that have been recently proposed reflect a better understanding of conidiogenesis.

MYCOLOGY

Mycology is the study of fungi, their relationships to each other and other organisms, and the unique biochemistry which sets them apart from other groups. Fungi are eukaryotic organisms which belongs to their own kingdom. Until advances in DNA technology, it was assumed that fungi were an offshoot of the plant kingdom. DNA and biochemical analysis has revealed that fungi are a separate lineage of eukaryotes, distinguished by their unique cell wall made of chitin and glucans which often surrounds multinucleated cells. Mycology is a necessary branch of biology because fungi are considerably different from both plants and animals.

Until the 1800's, it was assumed that fungi were simply a different kind of plant. Mushrooms, the reproductive bodies of fungi, were eaten, used as medicine, and used for their hallucinogenic effects since antiquity.

However, it was not until the advent of modern biochemistry and DNA analysis that it was fully realized how different fungi were. Instead of a cell wall made of cellulose, the wall in fungi is composed of glucans and chitin, molecules found in plants and insects, respectively. Instead of having a single nucleus, like most plants and animals, fungi are often multinucleated and contain special pores allowing the cytoplasm and nucleus to flow freely between various chambers in the fungal organism. DNA analysis revealed a closer relation to animals than plants. As scientists observed fungal lifecycles further, they realize that the majority of most fungi spends its time as a mold or ooze. This multicellular lifeform moves its way through decaying organic material, utilizing the minerals and organic molecules as it goes. Not only were fungi the major decomposing organism in the world, scientist also determined that certain fungi were responsible for events like fermentation and crop diseases.

With this, the field of mycology exploded. Agricultural mycology focuses on utilizing and controlling fungi in commercial crops. Toxicologists study mushroom and fungi for compounds which adversely affect other organisms. Pharmaceutical companies race to extract useful compounds from mushrooms. Careers in mycology are as diverse and complex as the field itself.

Mycology first became an important science in the agricultural industry and remains so today. A phytopathologist studies plant diseases, especially those which affect crops. Fungi are a major pest for many crops, but also serve symbiotic roles and allow plants to extract nutrients and water from the soil. Mycology is needed to distinguish between beneficial and harmful fungi, as well as to treat crops and prevent future infections. Further, certain types of fungi are used as pesticides, as they are more natural than synthetic pesticides and can kill targeted insects.

However, mycology has expanded well beyond its origins in agriculture. Once it was realized how broad and diverse the fungi kingdom is, the various roles of fungi in society were better understood. For instance, cheese is produced by various fungi. Mycology can classify and understand these organisms, leading to better and more efficiently produced cheese and dairy products. Yeast is also a form of fungi, and understanding the process of fermentation carried out by yeast is a science in itself. Fermentation science degrees can found from the bachelor level up, and graduates can work in the brewing and distilling industries, creating beer, wines and liquor. Yeast is also used in bread making, and microbiologists are required to maintain the cultures to produce enough yeast for bread production.

A specialized field of mycology is mycotoxicology, or the study of the toxins produced by mushrooms. Typically, a mycotoxicologists has a doctorate degree in biochemistry or organic chemistry, or a medical doctorate with concentrations in mycology and toxins. Fungi produce a variety of chemicals which have toxic effects on all kinds of organisms. Humans have eaten mushrooms since the earliest hunter-gatherers, but many mushrooms remain highly toxic. Other compounds found in mushrooms have potentially beneficial properties which could be used in medicine. Many mycotoxicologists work for pharmaceutical companies, trying to develop new drugs based on these compounds.

Mycology contains still more specializations, and is a continually evolving field. As more research is done, fungi are becoming a large and complex kingdom. Research is expanding and focusing on many special areas, including interesting applications for certain fungi. Some of these applications include radiotrophic fungi which appear to grow in the presence of radioactivity and could possibly alleviate radioactive wastes, and fungi which can break down complex organic substances into carbon dioxide. Many of these applications have tremendous commercial value, and researchers are needed at many institutions to explore these aspects of mycology.

Finally, an ethnomycologist is a scientist who studies the historical uses of fungi. Cultures have used mushroom as food, medicine, hallucinogens, and for a variety of other things. Ethnomycologists study these uses and inform the public and front-line researchers about which fungi have known effects and which are benign. Considering the immense size and diversity of fungi, and the relatively unorganized history of the classification of fungi, ethnomycologists provide a critical function in sorting through the dense but helpful information already gathered by past cultures and societies. The field of mycology is continually expanding as these many professions push the boundaries of knowledge and fill in the missing gaps.

Branches of Mycology

CHAPTER 2

Mycology can be divided into a number of branches to focus on specific areas of the field. Some of them are lichenology, mycotoxicology, ethnomycology, forensic mycology, etc. This chapter closely examines the varied branches of mycology to provide an extensive understanding of the subject.

LICHENOLOGY

The branch of mycology that studies lichens is known as lichenology. Lichens are organisms which interact symbiotically with fungi. The study of lichen borrows from many studies of science, including mycology, phycology, microbiology, and botany. A lichenologist is the name of scientist studying lichenology. It has origins going back to the 18th century, and has influenced many of the afore mentioned studies as much as it borrows from them.

While there is documentation from early times from people like Piliny the Elder or Theophrastus, the studies into lichen were not very deep. It wouldn't be until the 18th century when the father of lichenology, Joseph Pitton de Tournefort, would study them intensely and groped them into their own genus.

40 years after the works of Joseph Pitton de Tournefort, Dillenius would write a book Historia Muscorum. In it, he made the first divisions of the group created by Tournefort, which would create subfamilies Usnea, Coralloide, and Lichen, which was due to the morphological (shape, color, pattern, etc.) of lichen. Once this revolution in taxonomy, led by Linnaeus, created new systems in which to classify lichens. In the new system, Lichen would be divided into eight divisions according to how the lichens appeared, mainly the thallus of them. Thalli (plural of thallus) are vegetative tissue in lichen.

By the 20th century there were many advances in lichenology, and in 1909 Konstatntin Mereschkowski would present the research paper "The Theory of two Plasms as the basis of Symbiogenesis, A new study on the Origin of Organisms. Within it, he explained a new theory of symbiogensis (when life was first created), and often times referenced his previous work Nature and Origin of Chromatophores in the Plant Kingdom". All of this contributes to the Theory of Endosymbiosis, which is an evolutionary theory that focuses on the origins of eukaryotic cells (organisms with membranes) and prokaryotic organisms (unicellular organisms that lack membranes).

Lichenology has been used to help advance pharmacology (medicine) in many ways, but it wouldn't be until the 20th century when scientists would become interested in them. Various substances existed within lichen carry many antibacterial actions scientists would come to find. This would lead into many scientists studying mycology more in depth, as well as the worldwide implications of studying lichen growing every day. With more information into lichen, we will be able to help create new medicines as well as prolong the human race.

Lichen

A lichen is a composite organism that arises from algae or cyanobacteria living among filaments of multiple fungispecies in a mutualistic relationship. The combined lichen has properties different from those of its component organisms. Lichens come in many colors, sizes, and forms. The properties are sometimes plant-like, but lichens are not plants. Lichens may have tiny, leafless branches (fruticose), flat leaf-like structures (foliose), flakes that lie on the surface like peeling paint (crustose), a powder-like appearance (leprose), or other growth forms.

A tree covered with leafy foliose lichens and shrubby fruticose lichens.

A macrolichen is a lichen that is either bush-like or leafy; all other lichens are termed microlichens. Here, "macro" and "micro" do not refer to size, but to the growth form. Common names for lichens may contain the word moss (e.g., "reindeer moss", "Iceland moss"), and lichens may superficially look like and grow with mosses, but lichens are not related to mosses or any plant. Lichens do not have roots that absorb water and nutrients as plants do, but like plants, they produce their own nutrition by photosynthesis. When they grow on plants, they do not live as parasites, but instead use the plants as a substrate.

Letharia vulpina, wolf lichen, grows like a multiple-branched tuft or leafless mini-shrub, so it has a fruticose growth form.

Lichens occur from sea level to high alpine elevations, in many environmental conditions, and can grow on almost any surface. Lichens are abundant growing on bark, leaves, mosses, on other lichens, and hanging from branches "living on thin air" (epiphytes) in rain forests and in temperate woodland. They grow on rock, walls, gravestones, roofs, exposed soil surfaces, and in the soil as part of a biological soil crust. Different kinds of lichens have adapted to survive in some of the most extreme environments on Earth: arctic tundra, hot dry deserts, rocky coasts, and toxic slag heaps. They can even live inside solid rock, growing between the grains.

It is estimated that 6% of Earth's land surface is covered by lichens. There are about 20,000 known species of lichens. Some lichens have lost the ability to reproduce sexually, yet continue to speciate. Lichens can be seen as being relatively self-contained miniature ecosystems, where the fungi, algae, or cyanobacteria have the potential to engage with other microorganisms in a functioning system that may evolve as an even more complex composite organism.

Flavoparmelia caperata has leaf-like structures, so it is foliose.

Caloplaca marina grows like an orange crust coating the rock, so it is crustose.

Lichens may be long-lived, with some considered to be among the oldest living things. They are among the first living things to grow on fresh rock exposed after an event such as a landslide. The long life-span and slow and regular growth rate of some lichens can be used to date events (lichenometry).

Caloplaca thallincola [sv] grows like a crust, and in a pattern that radiates outward from the center, so it has a crustose placodioid growth form.

Pannaria lurida forms small leaf-like scales crustose below but free at the tips, so it is squamulose.

Growth Forms

Lichens grow in a wide range of shapes and forms (morphologies). The shape of a lichen is usually determined by the organization of the fungal filaments. The nonreproductive tissues, or vegetative body parts, are called the thallus. Lichens are grouped by thallus type, since the thallus is usually the most visually prominent part of the lichen. Thallus growth forms typically correspond to a few basic internal structure types. Common names for lichens often come from a growth form or color that is typical of a lichen genus.

Common groupings of lichen thallus growth forms are:

- Fruticose: Growing like a tuft or multiple-branched leafless mini-shrub, upright or hanging down, 3-dimensional branches with nearly round cross section (terete) or flattened.
- Foliose: Growing in 2-dimensional, flat, leaf-like lobes.
- Crustos: Crust-like, adhering tightly to a surface (substrate) like a thick coat of paint.
- Squamulose: Formed of small leaf-like scales crustose below but free at the tips.
- Leprose: Powdery.
- Gelatinous: Jelly-like.
- Filamentous: Stringy or like matted hair.
- Byssoid: Wispy, like teased wool.
- Structureless.

There are variations in growth types in a single lichen species, grey areas between the growth type descriptions, and overlapping between growth types.

When crustose lichen gets old, the center may start to crack up like old-dried paint, old-broken asphalt paving, or like the polygonal "islands" of cracked-up mud in a dried lakebed. This is called being rimose or areolate, and the "island" pieces separated by the cracks are called areolas. The areolas appear separated, but are (or were) connected by an underlying "prothallus" or "hypothallus". When a crustose lichen grows from a center and appears to radiate out, it is called crustose placodioid. When the edges of the areolas lift up from the substrate, it is called squamulose.

These growth form groups are not precisely defined. Foliose lichens may sometimes branch and appear to be fruticose. Fruticose lichens may have flattened branching parts and appear leafy. Squamulose lichens may appear where the edges lift up. Gelatinous lichens may appear leafy when dry. Means of telling them apart in these cases are in the sections below.

Structures involved in reproduction often appear as discs, bumps, or squiggly lines on the surface of the thallus. The thallus is not always the part of the lichen that is most visually noticeable. Some lichens can grow inside solid rock between the grains (endolithic lichens), with only the sexual fruiting part visible growing outside the rock. These may be dramatic in color or appearance. Forms of these sexual parts are not in the above growth form categories. The most visually noticeable reproductive parts are often circular, raised, plate-like or disc-like outgrowths, with crinkly edges.

Chrysothrix chlorina grows like powder dusted on the rock so it is leprose.

Collema nigrescens is gelatinous, without internal structure for its parts.

Color

Lichens come in many colors. Coloration is usually determined by the photosynthetic component. Special pigments, such as yellow usnic acid, give lichens a variety of colors, including reds, oranges, yellows, and browns, especially in exposed, dry habitats. In the absence of special pigments, lichens are usually bright green to olive gray when wet, gray or grayish-green to brown when dry. This is because moisture causes the surface skin (cortex) to become more transparent, exposing the green photobiont layer. Different colored lichens covering large areas of exposed rock surfaces, or lichens covering or hanging from bark can be a spectacular display when the patches of diverse colors "come to life" or "glow" in brilliant displays following rain.

Different colored lichens may inhabit different adjacent sections of a rock face, depending on the angle of exposure to light. Colonies of lichens may be spectacular in appearance, dominating much of the surface of the visual landscape in forests and natural places, such as the vertical "paint" covering the vast rock faces of Yosemite National Park.

Color is used in identification. Color changes depending on when a lichen is wet or dry. Color descriptions when used for identification are based on when the lichen is dry. Dry lichens with a cyanobacterium as the photosynthetic partner tend to be dark grey, brown, or black.

The underside of the leaf-like lobes of foliose lichens is a different color from the top side (dorsiventral), often brown or black, sometimes white. A fruticose lichen may have flattened "branches", appearing similar to a foliose lichen, but the underside of a leaf-like structure on a fruticose lichen is the same color as the top side. The leaf-like lobes of a foliose lichen may branch, giving the appearance of a fruticose lichen, but the underside will be a different color from the top side.

The sheen on some jelly-like gelatinous lichens is from mucilaginous secretions.

Internal Structure and Growth Forms

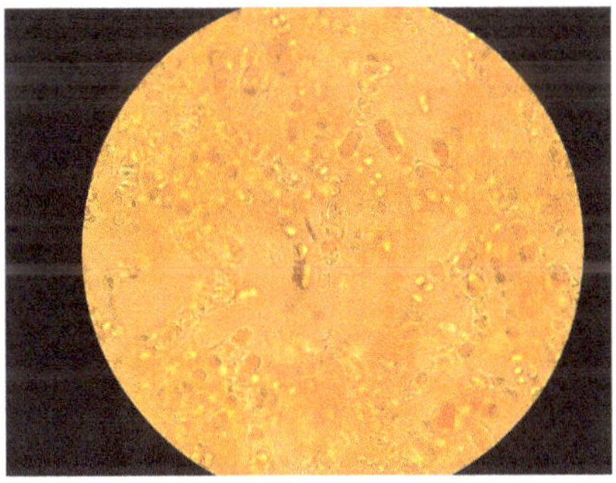

The cyanobacterium Hyella caespitosa with fungal hyphae in the lichen Pyrenocollema halodytes.

A lichen consists of a simple photosynthesizing organism, usually a green alga or cyanobacterium, surrounded by filaments of a fungus. Generally, most of a lichen's bulk is made of interwoven fungal filaments, although in filamentous and gelatinous lichens this is not the case. The fungus is called a mycobiont. The photosynthesizing organism is called a photobiont. Algal photobionts are called phycobionts. Cyanobacteria photobionts are called cyanobionts.

The part of a lichen that is not involved in reproduction, the "body" or "vegetative tissue" of a lichen, is called the thallus. The thallus form is very different from any form where the fungus or alga are growing separately. The thallus is made up of filaments of the fungus called hyphae. The filaments grow by branching then rejoining to create a mesh, which is called being "anastomose". The mesh of fungal filaments may be dense or loose.

Generally, the fungal mesh surrounds the algal or cyanobacterial cells, often enclosing them within complex fungal tissues that are unique to lichen associations. The thallus may or may not have a protective "skin" of densely packed fungal filaments, often containing a second fungal species, which is called a cortex. Fruticose lichens have one cortex layer wrapping around the "branches". Foliose lichens have an upper cortex on the top side of the "leaf", and a separate lower cortex on the bottom side. Crustose and squamulose lichens have only an upper cortex, with the "inside" of the lichen in direct contact with the surface they grow on (the substrate). Even if the edges peel up from the substrate and appear flat and leaf-like, they lack a lower cortex, unlike foliose lichens. Filamentous, byssoid, leprose, gelatinous, and other lichens do not have a cortex, which is called being ecorticate.

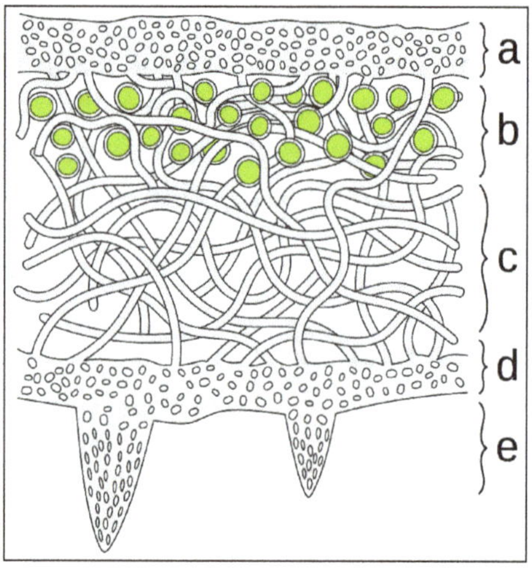

Schematic cross section of foliose lichen:

- The cortex is the outer layer of tightly woven fungus filaments (hyphae).
- This photobiont layer has photosynthesizing green algae.
- Loosely packed hyphae in the medulla.
- A tightly woven lower cortex.
- Anchoring hyphae called rhizineswhere the fungus attaches to the substrate.

Fruticose, foliose, crustose, and squamulose lichens generally have up to three different types of tissue, differentiated by having different densities of fungal filaments. The top layer, where the lichen contacts the environment, is called a cortex. The cortex is made of densely tightly woven, packed, and glued together (agglutinated) fungal filaments. The dense packing makes the cortex act like a protective "skin", keeping other organisms out, and reducing the intensity of sunlight on the layers below. The cortex layer can be up to several hundred micrometers (μm) in thickness (less than a millimeter). The cortex may be further topped by an epicortex of secretions, not cells, 0.6–1 μm thick in some lichens. This secretion layer may or may not have pores.

Below the cortex layer is a layer called the photobiontic layer or symbiont layer. The symbiont layer has less densely packed fungal filaments, with the photosynthetic partner embedded in them. The less dense packing allows air circulation during photosynthesis, similar to the anatomy of a leaf. Each cell or group of cells of the photobiont is usually individually wrapped by hyphae, and in some cases penetrated by a haustorium. In crustose and foliose lichens, algae in the photobiontic layer are diffuse among the fungal filaments, decreasing in gradation into the layer below. In fruticose lichens, the photobiontic layer is sharply distinct from the layer below.

The layer beneath the symbiont layer called is called the medulla. The medulla is less densely packed with fungal filaments than the layers above. In foliose lichens, there is usually, as in Peltigera, another densely packed layer of fungal filaments called the lower cortex. Root-like fungal structures called rhizines (usually) grow from the lower cortex to attach or anchor the lichen to the substrate. Fruticose lichens have a single cortex wrapping all the way around the "stems" and "branches". The medulla is the lowest layer, and may form a cottony white inner core for the branchlike thallus, or it may be hollow. Crustose and squamulose lichens lack a lower cortex, and the medulla is in direct contact with the substrate that the lichen grows on.

In crustose areolate lichens, the edges of the areolas peel up from the substrate and appear leafy. In squamulose lichens the part of the lichen thallus that is not attached to the substrate may also appear leafy. But these leafy parts lack a lower cortex, which distinguishes crustose and squamulose lichens from foliose lichens. Conversely, foliose lichens may appear flattened against the substrate like a crustose lichen, but most of the leaf-like lobes can be lifted up from the substrate because it is separated from it by a tightly packed lower cortex.

Gelatinous, byssoid, and leprose lichens lack a cortex (are ecorticate), and generally have only undifferentiated tissue, similar to only having a symbiont layer.

In lichens that include both green algal and cyanobacterial symbionts, the cyanobacteria may be held on the upper or lower surface in small pustules called cephalodia.

Pruinia is a whitish coating on top of an upper surface. An epinecral layer is "a layer of horny dead fungal hyphae with indistinct lumina in or near the cortex above the algal layer".

In August 2016, it was reported that macrolichens have more than one species of fungus in their tissues.

Physiology

Symbiotic Relation

A lichen is a composite organism that emerges from algae or cyanobacteria living among the filaments (hyphae) of the fungi in a mutually beneficial symbiotic relationship. The fungi benefit from the carbohydrates produced by the algae or cyanobacteria via photosynthesis. The algae or cyanobacteria benefit by being protected from the environment by the filaments of the fungi, which also gather moisture and nutrients from the environment, and (usually) provide an anchor to it. Although some photosynthetic partners in a lichen can survive outside the lichen, the lichen symbiotic association extends the ecological range of both partners, whereby most descriptions of lichen associations describe them as symbiotic. However, while symbiotic, the relationship is probably not mutualistic, since the algae give up a disproportionate amount of their sugars. Both partners gain water and mineral nutrients mainly from the atmosphere, through rain and dust. The fungal partner protects the alga by retaining water, serving as a larger capture area for mineral nutrients and, in some cases, provides minerals obtained from the substrate. If a cyanobacterium is present, as a primary partner or another symbiont in addition to a green alga as in certain tripartite lichens, they can fix atmospheric nitrogen, complementing the activities of the green alga.

In three different lineages the fungal partner has independently lost the mitochondrial gene atp9, which has key functions in mitochondrial energy production. The loss makes the fungi completely dependent on their symbionts.

The algal or cyanobacterial cells are photosynthetic and, as in plants, they reduce atmospheric carbon dioxide into organic carbon sugars to feed both symbionts. Phycobionts (algae) produce sugar alcohols (ribitol, sorbitol, and erythritol), which are absorbed by the mycobiont (fungus). Cyanobionts produce glucose. Lichenized fungal cells can make the photobiont "leak" out the products of photosynthesis, where they can then be absorbed by the fungus.

It appears many, probably the majority, of lichen also live in a symbiotic relationship with an order of basidiomycete yeasts called Cyphobasidiales. The absence of this third partner could explain the difficulties of growing lichen in the laboratory. The yeast cells are responsible for the formation of the characteristic cortex of the lichen thallus, and could also be important for its shape.

The lichen combination of alga or cyanobacterium with a fungus has a very different form (morphology), physiology, and biochemistry than the component fungus, alga, or cyanobacterium growing by itself, naturally or in culture. The body (thallus) of most lichens is different from those of either the fungus or alga growing separately. When grown in the laboratory in the absence of its photobiont, a lichen fungus develops as a structureless, undifferentiated mass of fungal filaments (hyphae). If combined with its photobiont under appropriate conditions, its characteristic form associated with the photobiont emerges, in the process called morphogenesis. In a few remarkable cases, a single lichen fungus can develop into two very different lichen forms when associating with either a green algal or a cyanobacterial symbiont. Quite naturally, these alternative forms were at first considered to be different species, until they were found growing in a conjoined manner.

Evidence that lichens are examples of successful symbiosis is the fact that lichens can be found in almost every habitat and geographic area on the planet. Two species in two genera of green algae are found in over 35% of all lichens, but can only rarely be found living on their own outside of a lichen.

In a case where one fungal partner simultaneously had two green algae partners that outperform each other in different climates, this might indicate having more than one photosynthetic partner at the same time might enable the lichen to exist in a wider range of habitats and geographic locations.

At least one form of lichen, the North American beard-like lichens, are constituted of not two but three symbiotic partners: an ascomycetous fungus, a photosynthetic alga, and, unexpectedly, a basidiomycetous yeast.

Phycobionts can have a net output of sugars with only water vapor. The thallus must be saturated with liquid water for cyanobionts to photosynthesize.

Algae produce sugars that are absorbed by the fungus by diffusion into special fungal hyphae called appressoria or haustoria in contact with the wall of the algal cells. The appressoria or haustoria may produce a substance that increases permeability of the algal cell walls, and may penetrate the walls. The algae may lose up to 80% of their sugar production to the fungus.

Ecology

Lichen associations may be examples of mutualism, commensalism or even parasitism, depending on the species. There is evidence to suggest that the lichen symbiosis is parasitic or commensalistic, rather than mutualistic. The photosynthetic partner can exist in nature independently of the fungal partner, but not vice versa. Photobiont cells are routinely destroyed in the course of nutrient exchange. The association is able to continue because reproduction of the photobiont cells matches the rate at which they are destroyed. The fungus surrounds the algal cells, often enclosing them within complex fungal tissues unique to lichen associations. In many species the fungus penetrates the algal cell wall, forming penetration pegs (haustoria) similar to those produced by pathogenic fungi that feed on a host. Cyanobacteria in laboratory settings can grow faster when they are alone rather than when they are part of a lichen.

Miniature Ecosystem and Holobiont Theory

Symbiosis in lichens is so well-balanced that lichens have been considered to be relatively self-contained miniature ecosystems in and of themselves. It is thought that lichens may be even more complex symbiotic systems that include non-photosynthetic bacterial communities performing other functions as partners in a holobiont.

Many lichens are very sensitive to environmental disturbances and can be used to cheaply assess air pollution, ozone depletion, and metal contamination. Lichens have been used in making dyes, perfumes, and in traditional medicines. A few lichen species are eaten by insects or larger animals, such as reindeer. Lichens are widely used as environmental indicators or bio-indicators. If air is very badly polluted with sulphur dioxide there may be no lichens present, just green algae may be found. If the air is clean, shrubby, hairy and leafy lichens become abundant. A few lichen species can tolerate quite high levels of pollution and are commonly found on pavements, walls and tree bark in urban areas. The most sensitive lichens are shrubby and leafy while the most tolerant lichens are all crusty in appearance. Since industrialisation many of the shrubby and leafy lichens such as Ramalina, Usnea and Lobaria species have very limited ranges, often being confined to the parts with the purest air.

Lichenicolous Fungi

Some fungi can only be found living on lichens as obligate parasites. These are referred to as lichenicolous fungi, and are a different species from the fungus living inside the lichen; thus they are not considered to be part of the lichen.

Reaction to Water

Moisture makes the cortex become more transparent. This way, the algae can conduct photosynthesis when moisture is available, and is protected at other times. When the cortex is more transparent, the algae show more clearly and the lichen looks greener.

Metabolites, Metabolite Structures and Bioactivity

Lichens can show intense antioxidant activity. Secondary metabolites are often deposited as crystals in the apoplast. Secondary metabolites are thought to play a role in preference for some substrates over others.

Sometimes lichens contain structures made from fungal metabolites, for example crustose lichens sometimes have a polysaccharide layer in the cortex.

Growth Rate

Lichens often have a regular but very slow growth rate of less than a millimeter per year.

In crustose lichens, the area along the margin is where the most active growth is taking place. Most crustose lichens grow only 1–2 mm in diameter per year.

Life Span

Lichens may be long-lived, with some considered to be among the oldest living organisms. Lifespan is difficult to measure because what defines the "same" individual lichen is not precise. Lichens grow by vegetatively breaking off a piece, which may or may not be defined as the "same" lichen, and two lichens can merge, then becoming the "same" lichen. An Arctic species called "map lichen" (Rhizocarpon geographicum) has been dated at 8,600 years, apparently the world's oldest living organism.

Response to Environmental Stress

Unlike simple dehydration in plants and animals, lichens may experience a complete loss of body water in dry periods. Lichens are capable of surviving extremely low levels of water content (poikilohydric). They quickly absorb water when it becomes available again, becoming soft and fleshy. Reconfiguration of membranes following a period of dehydration requires several minutes or more.

In tests, lichen survived and showed remarkable results on the adaptation capacity of photosynthetic activity within the simulation time of 34 days under Martian conditions in the Mars Simulation Laboratory (MSL) maintained by the German Aerospace Center (DLR).

The European Space Agency has discovered that lichens can survive unprotected in space. In an experiment led by Leopoldo Sancho from the Complutense University of Madrid, two species of lichen—Rhizocarpon geographicum and Xanthoria elegans—were sealed in a capsule and launched on a Russian Soyuz rocket 31 May 2005. Once in orbit, the capsules were opened and the lichens were directly exposed to the vacuum of space with its widely fluctuating temperatures and cosmic radiation. After 15 days, the lichens were brought back to earth and were found to be unchanged in their ability to photosynthesize.

Reproduction and Dispersal

Vegetative Reproduction

Xanthoparmelia sp. with dark-colored reproductive structures (disc-like apothecia) at center, surrounded by a pale coloured vegetative thallus.

Many lichens reproduce asexually, either by a piece breaking off and growing on its own (vegetative reproduction) or through the dispersal of diaspores containing a few algal cells surrounded by fungal cells. Because of the relative lack of differentiation in the thallus, the line between diaspore formation and vegetative reproduction is often blurred. Fruticose lichens can easily fragment, and new lichens can grow from the fragment (vegetative reproduction). Many lichens break up into fragments when they dry, dispersing themselves by wind action, to resume growth when moisture returns. Soredia (singular: "soredium") are small groups of algal cells surrounded by fungal filaments that form in structures called soralia, from which the soredia can be dispersed by wind. Isidia (singular: "isidium") are branched, spiny, elongated, outgrowths from the thallus that break off for mechanical dispersal. Lichen propagules (diaspores) typically contain cells from both partners, although the fungal components of so-called "fringe species" rely instead on algal cells dispersed by the "core species".

Sexual Reproduction

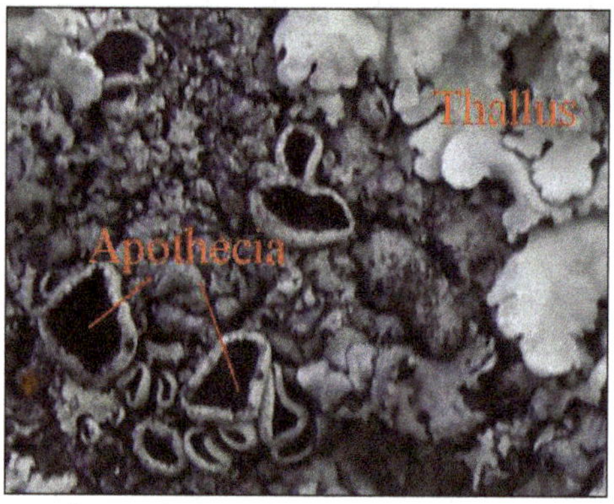

Disc-like apothecia (left) and thallus(right) on a foliose lichen.

Structures involved in reproduction often appear as discs, bumps, or squiggly lines on the surface of the thallus. Only the fungal partner in a lichen reproduces sexually. Many lichen fungi reproduce sexually like other fungi, producing spores formed by meiosis and fusion of gametes. Following dispersal, such fungal spores must meet with a compatible algal partner before a functional lichen can form.

Some lichen fungi belong to Basidiomycetes (basidiolichens) and produce mushroom-like reproductive structures resembling those of their nonlichenized relatives.

Most lichen fungi belong to Ascomycetes (ascolichens). Among the ascolichens, spores are produced in spore-producing structures called ascomata. The most common types of ascomata are the apothecium (plural: apothecia) and perithecium (plural: perithecia). Apothecia are usually cups or plate-like discs located on the top surface of the lichen thallus. When apothecia are shaped like squiggly line segments instead of like discs, they are called lirellae. Perithecia are shaped like flasks that are immersed in the lichen thallus tissue, which has a small hole for the spores to escape the flask, and appear like black dots on the lichen surface.

The three most common spore body types are raised discs called apothecia (singular: apothecium), bottle-like cups with a small hole at the top called perithecia (singular:

perithecium), and pycnidia (singular: pycnidium), shaped like perithecia but without asci (an ascus is the structure that contains and releases the sexual spores in fungi of the Ascomycota).

The apothecium has a layer of exposed spore-producing cells called asci (singular: ascus), and is usually a different color from the thallus tissue. When the apothecium has an outer margin, the margin is called the exciple. When the exciple has a color similar to colored thallus tissue the apothecium or lichen is called lecanorine, meaning similar to members of the genus Lecanora. When the exciple is blackened like carbon it is called lecideine meaning similar to members of the genus Lecidea. When the margin is pale or colorless it is called biatorine.

Crust-like thallus with pseudopodetia.

A "podetium" (plural: podetia) is a lichenized stalk-like structure of the fruiting body rising from the thallus, associated with some fungi that produce a fungal apothecium. Since it is part of the reproductive tissue, podetia are not considered part of the main body (thallus), but may be visually prominent. The podetium may be branched, and sometimes cup-like. They usually bear the fungal pycnidia or apothecia or both. Many lichens have apothecia that are visible to the naked eye.

Most lichens produce abundant sexual structures. Many species appear to disperse only by sexual spores. For example, the crustose lichens Graphis scripta and Ochrolechia parella produce no symbiotic vegetative propagules. Instead, the lichen-forming fungi of these species reproduce sexually by self-fertilization (i.e. they are homothallic). This breeding system may enable successful reproduction in harsh environments.

Mazaedia (singular: mazaedium) are apothecia shaped like a dressmaker's pin in (pin lichen)s, where the fruiting body is a brown or black mass of loose ascospores enclosed by a cup-shaped exciple, which sits on top of a tiny stalk.

Taxonomy and Classification

Lichens are classified by the fungal component. Lichen species are given the same scientific name (binomial name) as the fungus species in the lichen. Lichens are being integrated into the classification schemes for fungi. The alga bears its own scientific name, which bears no relationship to that of the lichen or fungus. There are about 13,500–17,000 identified lichen species. Nearly 20% of known fungal species are associated with lichens.

"Lichenized fungus" may refer to the entire lichen, or to just the fungus. This may cause confusion without context. A particular fungus species may form lichens with different algae species, giving rise to what appear to be different lichen species, but which are still classified as the same lichen species.

Formerly, some lichen taxonomists placed lichens in their own division, the Mycophycophyta, but this practice is no longer accepted because the components belong to separate lineages. Neither the ascolichens nor the basidiolichens form monophyletic lineages in their respective fungal phyla, but they do form several major solely or primarily lichen-forming groups within each phylum. Even more unusual than basidiolichens is the fungus Geosiphon pyriforme, a member of the Glomeromycota that is unique in that it encloses a cyanobacterial symbiont inside its cells. Geosiphon is not usually considered to be a lichen, and its peculiar symbiosis was not recognized for many years. The genus is more closely allied to endomycorrhizal genera. Fungi from Verrucariales also form marine lichens with the brown algae Petroderma maculiforme, and have a symbiotic relationship with seaweed like (rockweed) and Blidingia minima, where the algae are the dominant components. The fungi is thought to help the rockweeds to resist desiccation when exposed to air. In addition, lichens can also use yellow-green algae (Heterococcus) as their symbiotic partner.

Lichens independently emerged from fungi associating with algae and cyanobacteria multiple times throughout history.

Fungi

The fungal component of a lichen is called the mycobiont. The mycobiont may be an Ascomycete or Basidiomycete. The associated lichens are called either ascolichens or basidiolichens, respectively. Living as a symbiont in a lichen appears to be a successful way for a fungus to derive essential nutrients since about 20% of all fungal species have acquired this mode of life.

Thalli produced by a given fungal symbiont with its differing partners may be similar, and the secondary metabolites identical, indicating that the fungus has the dominant role in determining the morphology of the lichen. But the same mycobiont with different photobionts may also produce very different growth forms. Lichens are known in which there is one fungus associated with two or even three algal species.

Although each lichen thallus generally appears homogeneous, some evidence seems to suggest that the fungal component may consist of more than one genetic individual of that species. Two or more fungal species can interact to form the same lichen.

Photobionts

The photosynthetic partner in a lichen is called a photobiont. The photobionts in lichens come from a variety of simple prokaryotic and eukaryotic organisms. In the majority of lichens the photobiont is a green alga (Chlorophyta) or a cyanobacterium. In some lichens both types are present. Algal photobionts are called phycobionts, while cyanobacterial photobionts are called cyanobionts. According to one source, about 90% of all known lichens have phycobionts, and about 10% have cyanobionts, while another source states that two thirds of lichens have green algae as phycobiont, and about one third have a cyanobiont. Approximately 100 species of photosynthetic partners from 40 genera and five distinct classes (prokaryotic: Cyanophyceae; eukaryotic: Trebouxiophyceae, Phaeophyceae, Chlorophyceae) have been found to associate with the lichen-forming fungi.

Common algal photobionts are from the genera Trebouxia, Trentepohlia, Pseudotrebouxia, or Myrmecia. Trebouxia is the most common genus of green algae in lichens, occurring in about 40% of all lichens. "Trebouxioid" means either a photobiont that is in the genus Trebouxia, or resembles a member of that genus, and is therefore presumably a member of the class Trebouxiophyceae. The second most commonly represented green alga genus is Trentepohlia. Overall, about 100 species of eukaryotes are known to occur as photobionts in lichens. All the algae are probably able to exist independently in nature as well as in the lichen.

A "cyanolichen" is a lichen with a cyanobacterium as its main photosynthetic component (photobiont). Most cyanolichen are also ascolichens, but a few basidiolichen like Dictyonema and Acantholichen have cyanobacteria as their partner.

The most commonly occurring cyanobacterium genus is Nostoc. Other common cyanobacterium photobionts are from Scytonema. Many cyanolichens are small and black, and have limestone as the substrate. Another cyanolichen group, the jelly lichens of the genera Collema or Leptogium are gelatinous and live on moist soils. Another group of large and foliose species including Peltigera, Lobaria, and Degelia are grey-blue, especially when dampened or wet. Many of these characterize the Lobarion communities of higher rainfall areas in western Britain, e.g., in the Celtic rain forest. Strains of cyanobacteria found in various cyanolichens are often closely related to one another. They differ from the most closely related free-living strains.

The lichen association is a close symbiosis. It extends the ecological range of both partners but is not always obligatory for their growth and reproduction in natural environments, since many of the algal symbionts can live independently. A prominent example is the alga Trentepohlia, which forms orange-coloured populations on tree trunks and suitable rock faces. Lichen propagules (diaspores) typically contain cells from both partners, although the fungal components of so-called "fringe species" rely instead on algal cells dispersed by the "core species".

The same cyanobiont species can occur in association with different fungal species as lichen partners. The same phycobiont species can occur in association with different fungal species as lichen partners. More than one phycobiont may be present in a single thallus.

A single lichen may contain several algal genotypes. These multiple genotypes may better enable response to adaptation to environmental changes, and enable the lichen to inhabit a wider range of environments.

Controversy over Classification Method and Species Names

There are about 20,000 known lichen species. But what is meant by "species" is different from what is meant by biological species in plants, animals, or fungi, where being the same species implies that there is a common ancestral lineage. Because lichens are combinations of members of two or even three different biological kingdoms, these components must have a different ancestral lineage from each other. By convention, lichens are still called "species" anyway, and are classified according to the species of their fungus, not the species of the algae or cyanobacteria. Lichens are given the same scientific name (binomial name) as the fungus in them, which may cause some confusion. The alga bears its own scientific name, which has no relationship to the name of the lichen or fungus.

Depending on context, "lichenized fungus" may refer to the entire lichen, or to the fungus when it is in the lichen, which can be grown in culture in isolation from the algae or cyanobacteria. Some algae and cyanobacteria are found naturally living outside of the lichen. The fungal, algal, or cyanobacterial component of a lichen can be grown by itself in culture. When growing by themselves, the fungus, algae, or cyanobacteria have very different properties than those of the lichen. Lichen properties such as growth form, physiology, and biochemistry, are very different from the combination of the properties of the fungus and the algae or cyanobacteria.

The same fungus growing in combination with different algae or cyanobacteria, can produce lichens that are very different in most properties, meeting non-DNA criteria for being different "species". Historically, these different combinations were classified as different species. When the fungus is identified as being the same using modern DNA methods, these apparently different species get reclassified as the same species under the current convention for classification by fungal component. This has led to debate about this classification convention. These apparently different "species" have their own independent evolutionary history.

There is also debate as to the appropriateness of giving the same binomial name to the fungus, and to the lichen that combines that fungus with an alga or cyanobacterium (synecdoche). This is especially the case when combining the same fungus with different algae or cyanobacteria produces dramatically different lichen organisms, which would be considered different species by any measure other than the DNA of the fungal component. If the whole lichen produced by the same fungus growing in association with different algae or cyanobacteria, were to be classified as different "species", the number of "lichen species" would be greater.

Diversity

The largest number of lichenized fungi occur in the Ascomycota, with about 40% of species forming such an association. Some of these lichenized fungi occur in orders with nonlichenized fungi that live as saprotrophsor plant parasites (for example, the Leotiales, Dothideales, and Pezizales). Other lichen fungi occur in only five orders in which all members are engaged in this habit (Orders Graphidales, Gyalectales, Peltigerales, Pertusariales, and Teloschistales). Overall, about 98% of lichens have an ascomycetous mycobiont. Next to the Ascomycota, the largest number of lichenized fungi occur in the unassigned fungi imperfecti, a catch-all category for fungi whose sexual form of reproduction has never been observed. Comparatively few Basidiomycetes are lichenized, but these include agarics, such as species of Lichenomphalia, clavarioid fungi, such as species of Multiclavula, and corticioid fungi, such as species of Dictyonema.

Identification Methods

Lichen identification uses growth form and reactions to chemical tests. "Pd" refers to the outcome of the Pd test or is used as an abbreviation for the chemical used in the test, para-phenylenediamine. If putting a drop on a lichen turns an area bright yellow to orange, this helps identify it as belonging to either the genus Cladonia or Lecanora.

Evolution and Paleontology

The fossil record for lichens is poor. The extreme habitats that lichens dominate, such as tundra, mountains, and deserts, are not ordinarily conducive to producing fossils. There are fossilized lichens embedded in amber. The fossilized Anzia is found in pieces of amber in northern Europe and dates back approximately 40 million years. Lichen fragments are also found in fossil leaf beds, such as Lobaria from Trinity County in northern California, USA, dating back to the early to middle Miocene.

The oldest fossil lichen in which both symbiotic partners have been recovered is Winfrenatia,an early zygomycetous (Glomeromycotan) lichen symbiosis that may have involved controlled parasitism, is permineralized in the Rhynie Chert of Scotland, dating from early Early Devonian, about 400 million years ago. The slightly older fossil Spongiophyton has also been interpreted as a lichen on morphological and isotopic grounds, although the isotopic basis is decidedly shaky. It has been demonstrated that Silurian-Devonian fossils Nematothallus and Prototaxites were lichenized. Thus lichenized Ascomycota and Basidiomycota were a component of Early Silurian-Devonian terrestrial ecosystems.

Newer research suggests that lichen evolved after the evolution of land plants.

The ancestral ecological state of both Ascomycota and Basidiomycota was probably saprobism, and independent lichenization events may have occurred multiple times. In 1995, Gargas and colleagues proposed that there were at least five independent origins of lichenization; three in the basidiomycetes and at least two in the Ascomycetes. However, Lutzoni et al. indicate that lichenization probably evolved earlier and was followed by multiple independent losses. Some non-lichen-forming fungi may have secondarily lost the ability to form a lichen association. As a result, lichenization has been viewed as a highly successful nutritional strategy.

Lichenized Glomeromycota may extend well back into the Precambrian. Lichen-like fossils consisting of coccoid cells and thin filaments (mucoromycotinan Glomeromycota) are permineralized in marine phosphorite of the Doushantuo Formation in southern China. These fossils are thought to be 551 to 635 million years old or Ediacaran. Ediacaran acritarchs also have many similarities with Glomeromycotanvesicles and spores. It has also been claimed that Ediacaran fossils including Dickinsonia, were lichens, although this claim is controversial. Endosymbiotic Glomeromycota comparable with living Geosiphon may extend back into the Proterozoic in the form of 1500 million year old Horodyskia and 2200 million year old Diskagma. Discovery of these fossils suggest that fungi developed symbiotic partnerships with photoautotrophs long before the evolution of vascular plants.

Ecology and Interactions with Environment

Substrates and Habitats

Lichens on a statue made of limestone on a tower of Regensburg Cathedral.

Lichens cover about 7% of the planet's surface and grow on and in a wide range of substrates and habitats, including some of the most extreme conditions on earth. They are abundant growing on bark, leaves, and hanging from branches "living on thin air" (epiphytes) in rain forests and in temperate woodland. They grow on bare rock, walls, gravestones, roofs, and exposed soil surfaces. They can survive in some of the most extreme environments on Earth: arctic tundra, hot dry deserts, rocky coasts, and toxic slag heaps. They can live inside solid rock, growing between the grains, and in the soil as part of a biological soil crust in arid habitats such as deserts. Some lichens do not grow on anything, living out their lives blowing about the environment.

When growing on mineral surfaces, some lichens slowly decompose their substrate by chemically degrading and physically disrupting the minerals, contributing to the process of weathering by which rocks are gradually turned into soil. While this contribution to weathering is usually benign, it can cause problems for artificial stone structures. For example, there is an ongoing lichen growth problem on Mount Rushmore National Memorial that requires the employment of mountain-climbing conservators to clean the monument.

Lichens are not parasites on the plants they grow on, but only use them as a substrate to grow on. The fungi of some lichen species may "take over" the algae of other lichen species. Lichens make their own food from their photosynthetic parts and by absorbing minerals from the environment. Lichens growing on leaves may have the appearance of being parasites on the leaves, but they are not. However, some lichens, notably those of the genus Diploschistes are known to parasitise other lichens. Diploschistes muscorum starts its development in the tissue of a host Cladonia species.

In the arctic tundra, lichens, together with mosses and liverworts, make up the majority of the ground cover, which helps insulate the ground and may provide forage for grazing animals. An example is "Reindeer moss", which is a lichen, not a moss.

A crustose lichen that grows on rock is called a saxicolous lichen. Crustose lichens that grow on the rock are epilithic, and those that grow immersed inside rock, growing between the crystals with only their fruiting bodies exposed to the air, are called endolithic lichens. A crustose lichen that grows on bark is called a corticolous lichen. A lichen that grows on wood from which the bark has been stripped is called a lignicolous lichen. Lichens that grow immersed inside plant tissues are called endophloidic lichens or endophloidal lichens. Lichens that use leaves as substrates, whether the leaf is still on the tree or on the ground, are called epiphyllous or foliicolous. A terricolous lichen grows on the soil as a substrate. Many squamulous lichens are terricolous. Umbillicate lichens are foliose lichens that are attached to the substrate at only one point. A vagrant lichen is not attached to a substrate at all, and lives its life being blown around by the wind.

Lichens and Soils

In addition to distinct physical mechanisms by which lichens break down raw stone, recent studies indicate lichens attack stone chemically, entering newly chelated minerals into the ecology.

The lichen exudates, which have powerful chelating capacity, the widespread occurrence of mineral neoformation, particularly metal oxalates, together with the characteristics of weathered substrates, all confirm the significance of lichens as chemical weathering agents.

Over time, this activity creates new fertile soil from lifeless stone.

Lichens may be important in contributing nitrogen to soils in some deserts through being eaten, along with their rock substrate, by snails, which then defecate, putting the nitrogen into the soils. Lichens help bind and stabilize soil sand in dunes. In deserts and semi-arid areas, lichens are part of extensive, living biological soil crusts, essential for maintaining the soil structure. Lichens have a long fossil record in soils dating back 2.2 billion years.

Ecological Interactions

Lichens are pioneer species, among the first living things to grow on bare rock or areas denuded of life by a disaster. Lichens may have to compete with plants for access to sunlight, but because of their small size and slow growth, they thrive in places where higher plants have difficulty growing. Lichens are often the first to settle in places lacking soil, constituting the sole vegetation in some extreme environments such as those found at high mountain elevations and at high latitudes. Some survive in the tough conditions of deserts, and others on frozen soil of the Arctic regions.

A major ecophysiological advantage of lichens is that they are poikilohydric (poikilo- variable, hydric- relating to water), meaning that though they have little control over the status of their hydration, they can tolerate irregular and extended periods of severe desiccation. Like some mosses, liverworts, ferns, and a few "resurrection plants", upon desiccation, lichens enter a metabolic suspension or stasis (known as cryptobiosis) in which the cells of the lichen symbionts are dehydrated to a degree that halts most biochemical activity. In this cryptobiotic state, lichens can survive wider extremes of temperature, radiation and drought in the harsh environments they often inhabit.

Lichens suppress the growth of mosses and higher plants around them.

Lichens do not have roots and do not need to tap continuous reservoirs of water like most higher plants, thus they can grow in locations impossible for most plants, such as bare rock, sterile soil or sand, and various artificial structures such as walls, roofs and monuments. Many lichens also grow as epiphytes (epi- on the surface, phyte- plant) on plants, particularly on the trunks and branches of trees. When growing on plants, lichens are not parasites; they do not consume any part of the plant nor poison it. Lichens produce allelopathic chemicals that inhibit the growth of mosses. Some ground-dwelling lichens, such as members of the subgenus Cladina (reindeer lichens), produce allelopathic chemicals that leach into the soil and inhibit the germination of seeds, spruce and other plants. Stability (that is, longevity) of their substrate is a major factor of lichen habitats. Most lichens grow on stable rock surfaces or the bark of old trees, but many others grow on soil and sand. In these latter cases, lichens are often an important part of soil stabilization; indeed, in some desert ecosystems, vascular (higher) plant seeds cannot become established except in places where lichen crusts stabilize the sand and help retain water.

Lichens may be eaten by some animals, such as reindeer, living in arctic regions. The larvae of a number of Lepidoptera species feed exclusively on lichens. These include Common Footman and Marbled Beauty. However, lichens are very low in protein and high in carbohydrates, making them unsuitable for some animals. Lichens are also used by the Northern Flying Squirrel for nesting, food, and a water source during winter.

Effects of Air Pollution

If lichens are exposed to air pollutants at all times, without any deciduous parts, they are unable to avoid the accumulation of pollutants. Also lacking stomata and a cuticle, lichens may absorb aerosols and gases over the entire thallus surface from which they may readily diffuse to the photobiont layer. Because lichens do not possess roots, their primary source of most elements is the air, and therefore elemental levels in lichens often reflect the accumulated composition of ambient air. The processes by which atmospheric deposition occurs include fog and dew, gaseous absorption, and dry deposition. Consequently, many environmental studies with lichens emphasize their feasibility as effective biomonitors of atmospheric quality.

Some lichens, like the foliose Lobaria pulmonaria, are sensitive to air pollution.

Not all lichens are equally sensitive to air pollutants, so different lichen species show different levels of sensitivity to specific atmospheric pollutants. The sensitivity of a lichen to air pollution is directly related to the energy needs of the mycobiont, so that the stronger the dependency of the mycobiont on the photobiont, the more sensitive the lichen is to air pollution. Upon exposure to air pollution, the photobiont may use metabolic energy for repair of its cellular structures that would otherwise be used for maintenance of its photosynthetic activity, therefore leaving less metabolic energy available for the mycobiont. The alteration of the balance between the photobiont and mycobiont can lead to the breakdown of the symbiotic association. Therefore, lichen decline may result not only from the accumulation of toxic substances, but also from altered nutrient supplies that favor one symbiont over the other.

This interaction between lichens and air pollution has been used as a means of monitoring air quality since 1859, with more systematic methods developed by William Nylander in 1866.

Human Use

Food

Lichens are eaten by many different cultures across the world. Although some lichens are only eaten in times of famine, others are a staple food or even a delicacy. Two obstacles are often encountered when eating lichens: lichen polysaccharides are generally indigestible to humans, and lichens usually contain mildly toxic secondary compounds that should be removed before eating. Very few lichens are poisonous, but those high in vulpinic acid or usnic acid are toxic. Most poisonous lichens are yellow.

Iwatake (Umbilicaria esculenta) gathering at Kumano in Kishū, by Hiroshige II.

In the past, Iceland moss was an important source of food for humans in northern Europe, and was cooked as a bread, porridge, pudding, soup, or salad. Wila (Bryoria fremontii) was an important food in parts of North America, where it was usually pit-cooked. Northern peoples in North America and Siberia traditionally eat the partially digested reindeer lichen (Cladina spp.) after they remove it from the rumen of caribou or reindeer that have been killed. Rock tripe (Umbilicaria spp. and Lasalia spp.) is a lichen that has frequently been used as an emergency food in North America, and one species, Umbilicaria esculenta, is used in a variety of traditional Korean and Japanese foods.

Lichenometry

Lichenometry is a technique used to determine the age of exposed rock surfaces based on the size of lichen thalli. Introduced by Beschel in the 1950s, the technique has found many applications. It is used in archaeology, palaeontology, and geomorphology. It uses the presumed regular but slow rate of lichen growth to determine the age of exposed rock. Measuring the diameter (or other size measurement) of the largest lichen of a species on a rock surface indicates the length of time since the rock surface was first exposed. Lichen can be preserved on old rock faces for up to 10,000 years, providing the maximum age limit of the technique, though it is most accurate (within 10% error) when applied to surfaces that have been exposed for less than 1,000 years. Lichenometry is especially useful for dating surfaces less than 500 years old, as radiocarbon dating techniques are less accurate over this period. The lichens most commonly used for lichenometry are those of the genera Rhizocarpon (e.g. the species Rhizocarpon geographicum) and Xanthoria.

Biodegradation

Lichens have been shown to degrade polyester resins, as can be seen in archaeological sites in the Roman city of Baelo Claudia in Spain. Lichens can accumulate several environmental pollutants such as lead, copper, and radionuclides.

As Dyes

Many lichens produce secondary compounds, including pigments that reduce harmful amounts of sunlight and powerful toxins that reduce herbivory or kill bacteria. These compounds are very useful for lichen identification, and have had economic importance as dyes such as cudbear or primitive antibiotics.

The pH indicator (indicated acidic or basic) in the litmus test is a dye extracted from the lichen Roccella tinctoria by boiling.

In the Highlands of Scotland, traditional dyes for Harris tweed and other traditional cloths were made from lichens, including the orange Xanthoria parietina and the grey foliaceous Parmelia saxatilis common on rocks known as “crottle”.

There are reports dating almost 2,000 years old of lichens being used to make purple and red dyes. Of great historical and commercial significance are lichens belonging to the family Roccellaceae, commonly called orchella weed or orchil. Orcein and other lichen dyes have largely been replaced by synthetic versions.

Traditional Medicine and Research

Historically in traditional medicine of Europe, Lobaria pulmonaria was collected in large quantities as "Lungwort", due to its lung-like appearance (the doctrine of signatures suggesting that herbs can treat body parts that they physically resemble). Similarly, Peltigera leucophlebia was used as a supposed cure for thrush, due to the resemblance of its cephalodia to the appearance of the disease.

Lichens produce metabolites in research for their potential therapeutic or diagnostic value. Some metabolites produced by lichens are structurally and functionally similar to broad-spectrum antibiotics while few are associated respectively to antiseptic similarities. Usnic acid is the most commonly studied metabolite produced by lichens. It is also under research as an bactericidal agent against Escherichia coli and Staphylococcus aureus.

Aesthetic Appeal

Colonies of lichens may be spectacular in appearance, dominating the surface of the visual landscape as part of the aesthetic appeal to visitors of Yosemite National Park and Sequoia National Park. Orange and yellow lichens add to the ambience of desert trees, rock faces, tundras, and rocky seashores. Intricate webs of lichens hanging from tree branches add a mysterious aspect to forests. Fruticose lichens are used in model railroading and other modeling hobbies as a material for making miniature trees and shrubs.

Pine forest with Cladonia lichen ground-cover.

In Literature

In early Midrashic literature, the Hebrew word "vayilafeth" in Ruth 3:8 is explained as referring to Ruth entwining herself around Boaz like lichen. The tenth century Arab physician, Al-Tamimi, mentions lichens dissolved in vinegar and rose water being used in his day for the treatment of skin diseases and rashes.

The plot of John Wyndham's novel Trouble with Lichen revolves around an anti-aging chemical extracted from a lichen.

MYCOTOXICOLOGY

Mycotoxicology is the branch of mycology that focuses on analyzing and studying the toxins produced by fungi, known as mycotoxins. In the food industry it is important to adopt measures that keep mycotoxin levels as low as practicable, especially those that are heat-stable. These chemical compounds are the result of secondary metabolism initiated in response to specific developmental or environmental signals. This includes biological stress from the environment, such as lower nutrients or competition for those available. Under this secondary path the fungus produces a wide array of compounds in order to gain some level of advantage, such as incrementing the efficiency of metabolic processes to gain more energy from less food, or attacking other microorganisms and being able to use their remains as a food source.

Mycotoxins are made by fungi and are toxic to vertebrates and other animal groups in low concentrations. Low-molecular-weight fungal metabolites such as ethanol that are toxic only in high concentrations are not considered mycotoxins. Mushroom poisons are fungal metabolites that can cause disease and death in humans and other animals; they are rather arbitrarily excluded from discussions of mycotoxicology. Molds make mycotoxins; mushrooms and other macroscopic fungi make mushroom poisons. The distinction between a mycotoxin and a mushroom poison is based not only on the size of the producing fungus, but also on human intention. Mycotoxin exposure is almost always accidental. In contrast, with the exception of the victims of a few mycologically accomplished murderers, mushroom poisons are usually ingested by amateur mushroom hunters who have collected, cooked, and eaten what was misidentified as a harmless, edible species.

Mycotoxins are hard to define and are also very difficult to classify. Mycotoxins have diverse chemical structures, biosynthetic origins, myriad biological effects, and produce numerous different fungal species. Classification generally reflects the training of the categorizer and does not adhere to any set system. Mycotoxins are often arranged by physicians depending on what organ they effect. Mycotoxins can be categorized as nephrotoxins, hepatoxins, immunotoxins, neurotoxins, etc. Generic groups created by cell biologist are teratogens, mutagens, allergens, and carcinogens. Organic chemists have attempted to classify them by their chemical structures (e.g., lactones, coumarins); biochemists according to their biosynthetic origins (polyketides, amino acid-derived, etc.); physicians by the illnesses they cause (e.g., St. Anthony's fire, stachybotryotoxicosis), and mycologists by the fungi that produce them (e.g., Aspergillus toxins, Penicillium toxins). None of these classifications is entirely satisfactory. Aflatoxin, for example, is a hepatotoxic, mutagenic, carcinogenic, difuran-containing, polyketide-derived

Aspergillus toxin. Zearalenone is a Fusarium metabolite with potent estrogenic activity; hence, in addition to being called (probably erroneously) a mycotoxin, it also has been labeled a phytoestrogen, a mycoestrogen, and a growth promotant.

Aflatoxins

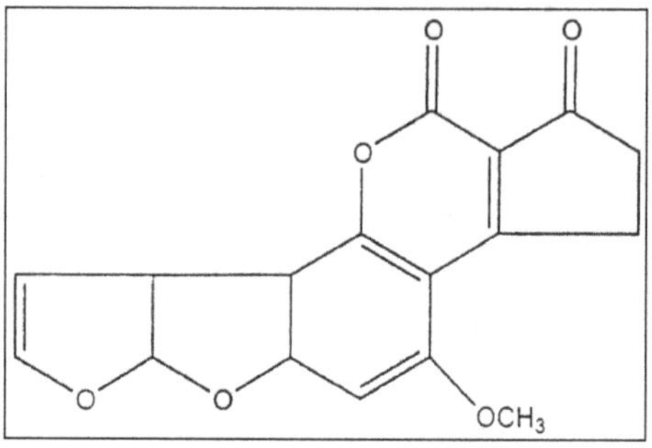

The aflatoxins were isolated and characterized after the death of more than 100,000 turkey poults (turkey X disease) was traced to the consumption of a mold-contaminated peanut meal. The four major aflatoxins are called B_1, B_2, G_1, and G_2 based on their fluorescence under UV light (blue or green) and relative chromatographic mobility during thin-layer chromatography. Aflatoxin B_1 is the most potent natural carcinogen known and is usually the major aflatoxin produced by toxigenic strains.

Aflatoxin B_1

Aflatoxins are difuranocoumarin derivatives produced by a polyketide pathway by many strains of Aspergillus flavus and Aspergillus parasiticus; in particular, Aspergillus flavus is a common contaminant in agriculture. Aspergillus bombycis, Aspergillus ochraceoroseus, Aspergillus nomius, and Aspergillus pseudotamari are also aflatoxin-producing species, but they are encountered less frequently. From the mycological perspective, there are great qualitative and quantitative differences in the toxigenic abilities displayed by different strains within each aflatoxigenic species. For example, only about half of Aspergillus flavus strains produce aflatoxins, while those that do may produce more than 10^6 µg/kg.

Many substrates support growth and aflatoxin production by aflatoxigenic molds. Natural contamination of cereals, figs, oilseeds, nuts, tobacco, and a long list of other commodities is a common occurrence. Like the genetic ability to make aflatoxin, contamination is highly variable. Sometimes crops become contaminated with aflatoxin in the field before harvest, where it is usually associated with drought stress; even more problematic is the fate of crops stored under conditions that favor mold growth. In storage, usually the most important variables are the moisture content of the substrate and the relative humidity of the surroundings. Aflatoxin contamination has been linked to increased mortality in farm animals and thus significantly lowers the value of grains as an animal feed and as an export commodity. Milk products can also serve as an indirect source of aflatoxin. When cows consume aflatoxin-contaminated feeds, they metabolically biotransform aflatoxin B_1 into a hydroxylated form called aflatoxin M_1.

Aflatoxin is associated with both toxicity and carcinogenicity in human and animal populations. The diseases caused by aflatoxin consumption are loosely called aflatoxicoses. Acute aflatoxicosis results in death; chronic aflatoxicosis results in cancer, immune suppression, and other "slow" pathological conditions. The liver is the primary target organ, with liver damage occurring when poultry, fish, rodents, and nonhuman primates are fed aflatoxin B_1. There are substantial differences in species susceptibility. Moreover, within a given species, the magnitude of the response is influenced by age, sex, weight, diet, exposure to infectious agents, and the presence of other mycotoxins and pharmacologically active substances. Thousands of studies on aflatoxin toxicity have been conducted, mostly concerning laboratory models or agriculturally important species.

Cytochrome P450 enzymes convert aflatoxins to the reactive 8,9-epoxide form, which is capable of binding to both DNA and proteins. Mechanistically, it is known that the reactive aflatoxin epoxide binds to the N^7 position of guanines. Moreover, aflatoxin B_1-DNA adducts can result in GC to TA transversions. A reactive glutathione S-transferase system found in the cytosol and microsomes catalyzes the conjugation of activated aflatoxins with reduced glutathione, leading to the excretion of aflatoxin. Variations in the level of the glutathione transferase system as well as variations in the cytochrome P450 system are thought to contribute to the differences observed in interspecies aflatoxin susceptibility.

Because of the differences in aflatoxin susceptibility in test animals, it has been difficult to extrapolate the possible effects of aflatoxin to humans, but acute toxicity of aflatoxins in Homo sapiens has not been observed very often.

It has been hypothesized that kwashiorkor, a severe malnutrition disease, may be a form of pediatric aflatoxicosis. Further early speculations that aflatoxin might be involved in Reye's syndrome, an encephalopathy, and fatty degeneration of the viscera in children and adolescents have not been substantiated. Nevertheless, aflatoxin has achieved some notoriety as a poison. The plot of The Human Factor, a spy thriller by Graham Greene, revolves around the murder of a central figure whose whiskey was laced with aflatoxin (a toxicologically improbable way to kill someone). Nevertheless, aflatoxin's reputation has a potent poison may explain why it has been adopted for use in bioterrorism. There is substantial evidence that Iraq stockpiled aflatoxin to be delivered in missiles.

The data on aflatoxin as a human carcinogen are far more damning than the data implicating it in acute human toxicities. Exposure to aflatoxins in the diet is considered an important risk factor for the development of primary hepatocellular carcinoma, particularly in individuals already exposed to hepatitis B. In classical epidemiology, several studies have linked liver cancer incidence to estimated aflatoxin consumption in the diet. In molecular epidemiology, it is possible to demonstrate with more certainty the association between putative carcinogens and specific cancers. Biomonitoring of aflatoxins can be done by analyzing for the presence of aflatoxin metabolites in blood, milk, and urine; moreover, excreted DNA adducts and blood protein adducts can also be monitored. The aflatoxin B_1-N^7-guanine adduct represents the most reliable urinary biomarker for aflatoxin exposure but reflects only recent exposure. Numerous studies have shown that carcinogenic potency is highly correlated with the extent of total DNA adducts formed in vivo.

Inactivation of the p53 tumor suppressor gene may be important in the development of primary hepatocellular carcinoma. Studies of liver cancer patients in Africa and China have shown that a mutation in the p53 tumor suppressor gene at codon 249 is associated with a G-to-T transversion. Mechanistically, it is known that the reactive aflatoxin epoxide binds to the N^7 position of guanines. Moreover, aflatoxin B_1-DNA adducts can result in GC to TA transversions. The specific mutation in codon 249 of the p53 gene has been called the first example of a "carcinogen-specific" biomarker that remains fixed in the tumor tissue.

There is also considerable evidence associating aflatoxin with neoplasms in extra-hepatic tissues, particularly the lungs. For example, one early epidemiological study of Dutch peanut processing workers exposed to dust contaminated with aflatoxin B_1 showed a correlation between both respiratory cancer and total cancer in the exposed group compared with unexposed cohorts. Anecdotal and circumstantial evidence of air-borne aflatoxin exposure leading to cancer has been reported. Deger concluded that dust from scrapings of chromatographic plates from aflatoxin analyses contributed to causing cancer in two young adults. Aflatoxin is a pulmonary carcinogen in experimental animals. The nonhepatic effects of aflatoxin B_1 have been summarized by Coulombe.

To recapitulate, there is no other natural product for which the data on human carcinogenicity are so compelling.

In developed countries, sufficient amounts of food combined with regulations that monitor aflatoxin levels in these foods protect human populations from significant aflatoxin ingestion. However, in countries where populations are facing starvation or where regulations are either not enforced or nonexistent, routine ingestion of aflatoxin may occur. Worldwide, liver cancer incidence rates are 2 to 10 times higher in developing countries than in developed countries. Unfortunately, strict limitation of aflatoxin-contaminated food is not always an option. A joint Food and Agriculture Organization/World Health Organization/United Nations Environment Programme Conference report stated that "in developing countries, where food supplies are already limited, drastic legal measure may lead to lack of food and to excessive prices. It must be remembered that people living in these countries cannot exercise the option of starving to death today in order to live a better life tomorrow".

There has been considerable fundamental work on the aflatoxins, especially on their biosynthesis and molecular biology. The first stable step in the biosynthetic pathway is the production of norsolorinic acid, an anthraquinone precursor, by a type II polyketide synthase. An elaborate series of at least 15 post-polyketide synthase steps follows, yielding a series of increasingly toxigenic metabolites. Sterigmatocystin, a related dihydrofuran toxin, is a late metabolite in the aflatoxin pathway and is also produced as a final biosynthetic product by a number of species such as Aspergillus versicolor and Aspergillus nidulans. Sterigmatocystin is both mutagenic and tumorigenic but is less potent than aflatoxin. Analysis of the molecular genetics of sterigmatocystin biosynthesis in the genetically tractable species Aspergillus nidulans has provided a useful model system. The genes for the sterigmatocystin gene cluster from Aspergillus nidulans have been cloned and sequenced. Cognate genes for aflatoxin pathway enzymes from Aspergillus flavus and Aspergillus parasiticus show high sequence similarity to the sterigmatocystin pathway genes. The organization of genes in the Aspergillus flavus, Aspergillus nidulans, and Aspergillus parasiticus sterigmatocystin-aflatoxin pathway has been compared by Cary et al. and Hicks et al.

Finally, it should be mentioned that Aspergillus oryzae and Aspergillus sojae, species that are widely used in Asian food fermentations such as soy sauce, miso, and sake, are closely related to the aflatoxigenic species Aspergillus flavus and Aspergillus parasiticus. Although these food fungi have never been shown to produce aflatoxin, they contain homologues of several aflatoxin biosynthesis pathway genes. Deletions and other genetic defects have led to silencing of the aflatoxin pathway in both Aspergillus oryzae and Aspergillus sojae.

Citrinin

Citrinin was first isolated from Penicillium citrinum prior to World War II; subsequently, it was identified in over a dozen species of Penicillium and several species of Aspergillus , including certain strains of Penicillium camemberti (used to produce cheese) and Aspergillus oryzae (used to produce sake, miso, and soy sauce). More recently, citrinin has also been isolated from Monascus ruber and Monascus purpureus, industrial species used to produce red pigments.

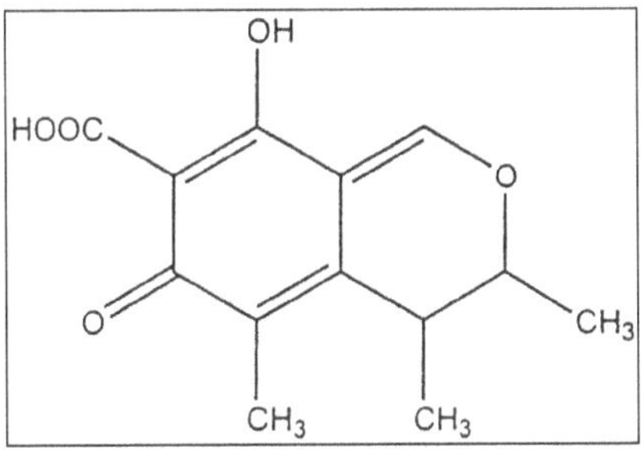

Citrinin has been associated with yellow rice disease in Japan. It has also been implicated as a contributor to porcine nephropathy. Citrinin acts as a nephrotoxin in all animal species tested, but its acute toxicity varies in different species. The 50% lethal dose for ducks is 57 mg/kg; for chickens it is 95 mg/kg; and for rabbits it is 134 mg/kg. Citrinin can act synergistically with ochratoxin A to depress RNA synthesis in murine kidneys.

Wheat, oats, rye, corn, barley, and rice have all been reported to contain citrinin. With immunoassays, citrinin was detected in certain vegetarian foods colored with Monascus pigments. Citrinin has also been found in naturally fermented sausages from Italy. Although citrinin is regularly associated with human foods, its significance for human health is unknown.

Ergot Alkaloids

The ergot alkaloids are among the most fascinating of fungal metabolites. They are classified as indole alkaloids and are derived from a tetracyclic ergoline ring system. Lysergic acid, a structure common to all ergot alkaloids, was first isolated in 1934. The clavines have ergoline as a basic structure but lack peptide components; the lysergic acid alkaloids include ergotamine and lysergic acid amide.

The structure of ergotamine:

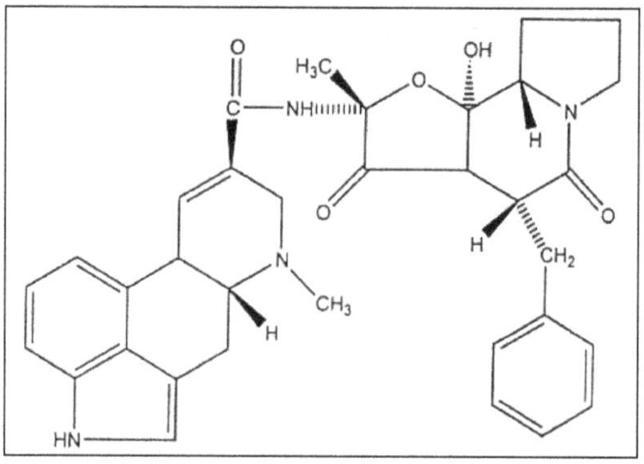

These compounds are produced as a toxic cocktail of alkaloids in the sclerotia of species of Claviceps, which are common pathogens of various grass species. The ingestion of these sclerotia, or ergots, has been associated with diseases since antiquity. An Assyrian tablet dated to 600 B.C.E., referring to a "noxious pustule in the ear of grain," is believed to be an early reference to ergot. The human disease acquired by eating cereals infected with ergot sclerotia, usually in the form of bread made from contaminated flour, is called ergotism or St. Anthony's fire. Two forms of ergotism are usually recognized, gangrenous and convulsive. The gangrenous form affects the blood supply to the extremities, while convulsive ergotism affects the central nervous system.

Human ergotism was common in Europe in the Middle Ages. For example, a three-volume work entitled Handbook of Geographical and Historical Pathology published in London by August Hirch between 1883 and 1886 recorded 132 epidemics of European ergotism between the 6th and 18th centuries. Matossian has suggested that the "slow nervous fever" described by the 18th century English physician Jon Huxham may be another example of human ergotism. Slow nervous fever usually occurred in the summer and fall after a severe winter; Huxham suspected "bad food" as the source of the trouble. Matossian has also postulated that ergot alkaloids may have had a strong influence on fertility trends in

England and other European countries during the 17th and 18th centuries. Modern methods of grain cleaning have almost eliminated ergotism as a human disease.

Sometimes the line between toxin and drug is defined with the shift of a decimal point or a change in a small chemical moiety. The ergot alkaloids are a case in point. Their myriad actions have long engaged the interest of physicians and pharmacologists. Several ergot alkaloids induce smooth muscle contractions. For centuries it had been observed that grazing on grass infected with ergot caused abortion in pregnant farm animals, so it is not surprising that midwives and others adopted ergot as a folk medicine, using it as both an abortifacient and a drug to accelerate to uterine contractions for women in labor.

Fumonisins

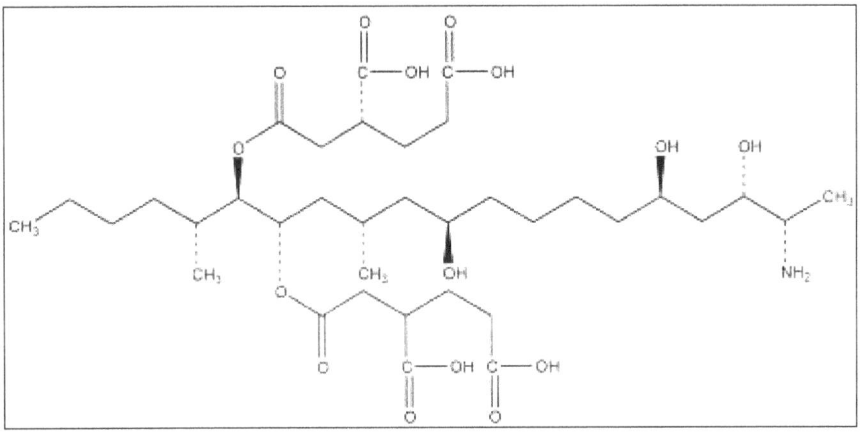

Fumonisins were first described and characterized in 1988. The most abundantly produced member of the family is fumonisin B_1. They are thought to be synthesized by condensation of the amino acid alanine into an acetate-derived precursor. Fumonisins are produced by a number of Fusarium species, notably Fusarium verticillioides (formerly Fusarium moniliforme = Gibberella fujikuroi), Fusarium proliferatum, and Fusarium nygamai, as well as Alternaria alternata f. sp. lycopersici. These fungi are taxonomically challenging, with a complex and rapidly changing nomenclature which has perplexed many nonmycologists (and some mycologists, too). The major species of economic importance is Fusarium verticillioides, which grows as a corn endophyte in both vegetative and reproductive tissues, often without causing disease symptoms in the plant. However, when weather conditions, insect damage, and the appropriate fungal and plant genotype are present, it can cause seedling blight, stalk rot, and ear rot. Fusarium verticillioides is present in virtually all corn samples. Most strains do not produce the toxin, so the presence of the fungus does not necessarily mean that fumonisin is also present. Although it is phytotoxic, fumonisin B_1 is not required for plant pathogenesis.

Fumonisin B_1

Fumonisins affect animals in different ways by interfering with sphingolipid metabolism.

They cause leukoencephalomalacia (hole in the head syndrome) in equines and rabbits; pulmonary edema and hydrothorax in swine; and hepatotoxic and carcinogenic effects and apoptosis in the liver of rats. In humans, there is a probable link with esophageal cancer. The occurrence of fumonisin B_1 is correlated with the occurrence of a higher incidence of esophageal cancer in regions of Transkei, China, and northeast Italy. It has been isolated at high levels in corn meal and corn grits, including seven samples from a supermarket in Charleston, S.C., a city which has the highest incidence of esophageal cancer among African-Americans in the United States. Several other mycotoxins, nutritional parameters, and other factors have also been implicated in the etiology of human esophageal cancer.

Unlike most known mycotoxins, which are soluble in organic solvents, fumonisins are hydrophilic. This makes them difficult to study. Usually they are extracted in aqueous methanol or aqueous acetonitrile. High-performance liquid chromatography with fluorescent detection is the most widely used analytic method. The fumonisin story raises the specter that there may be many other occult but toxic products of fungal metabolism that have not yet been discovered because of their hydrophilic nature.

Ochratoxin

Ochratoxin A was discovered as a metabolite of Aspergillus ochraceus in 1965 during a large screen of fungal metabolites that was designed specifically to identify new mycotoxins. Shortly thereafter, it was isolated from a commercial corn sample in the United States and recognized as a potent nephrotoxin. Members of the ochratoxin family have been found as metabolites of many different species of Aspergillus, including Aspergillus alliaceus, Aspergillus auricomus, Aspergillus carbonarius, Aspergillus glaucus, Aspergillus melleus, and Aspergillus niger. Because Aspergillus niger is used widely in the production of enzymes and citric acid for human consumption, it is important to ensure that industrial strains are nonproducers. Although some early reports implicated several Penicillium species, it is now thought that Penicillium verrucosum, a common contaminant of barley, is the only confirmed ochratoxin producer in this genus.

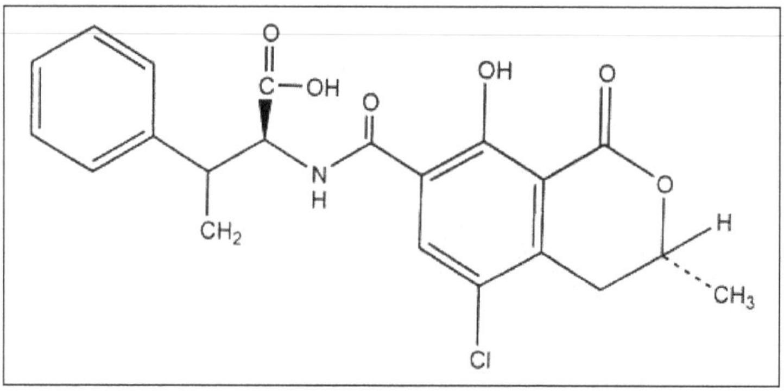

Ochratoxin A.

As with other mycotoxins, the substrate on which the molds grow as well as the moisture level, temperature, and presence of competitive microflora interact to influence the level of toxin produced. Ochratoxin A has been found in barley, oats, rye, wheat, coffee beans, and other plant products, with barley having a particularly high likelihood of contamination. There is also concern that ochratoxin may be present in certain wines, especially those from grapes contaminated with Aspergillus carbonarius.

Of the Aspergillus toxins, only ochratoxin is potentially as important as the aflatoxins. The kidney is the primary target organ. Ochratoxin A is a nephrotoxin to all animal species studied to date and is most likely toxic to humans, who have the longest half-life for its elimination of any of the species examined. In addition to being a nephrotoxin, animal studies indicate that ochratoxin A is a liver toxin, an immune suppressant, a potent teratogen, and a carcinogen. Ochratoxin A disturbs cellular physiology in multiple ways, but it seems that the primary effects are associated with the enzymes involved in phenylalanine metabolism, mostly by inhibiting the enzyme involved in the synthesis of the phenylalanine-tRNA complex. In addition, it inhibits mitochondrial ATP production and stimulates lipid peroxidation.

Ochratoxin has been detected in blood and other animal tissues and in milk, including human milk. It is frequently found in pork intended for human consumption. Ochratoxin is believed to be responsible for a porcine nephropathy that has been studied intensively in the Scandinavian countries. The disease is endemic in Denmark, where rates of porcine nephropathy and ochratoxin contamination in pig feed are highly correlated. In addition, ochratoxin is associated with disease and death in poultry.

There has been speculation that ochratoxins are involved in a human disease called endemic Balkan nephropathy. This condition is a progressive chronic nephritis that occurs in populations who live in areas bordering the Danube River in parts of Romania, Bulgaria, and the former Yugoslavia. In one Bulgarian study, ochratoxin contamination of food and the presence of ochratoxin in human serum were more common in families with endemic Balkan nephropathy and urinary tract tumors than in unaffected families. In addition to ochratoxin poisoning, this curious disease has been attributed to genetic factors, heavy metals, and possible occult infectious agents. The current consensus is that endemic Balkan nephropathy is of unknown etiology, but many mycotoxin reviews list it, without caveat, as an ochratoxicosis.

It has also been hypothesized that the gene for phenylketonuria might occur in relatively high frequency because of a heterozygous advantage against ochratoxin poisoning and that ochratoxin might be a risk factor for testicular cancer.

How common is human exposure to ochratoxin? Studies from Canada, Sweden, West Germany, and Yugoslavia detected ochratoxin in human blood and serum.

Several detailed risk assessments have been conducted for ochratoxin A. Given the known human exposure and the abundance of toxicological data from animal studies, the European Union Scientific Committee has recommended that ochratoxin A levels be reduced to below 5 ng/kg of body weight per day. In addition, several

European countries have proposed individual regulations, with maximum tolerated concentrations varying greatly from country to country. The International Agency for Research on Cancer has rated ochratoxin as a possible human carcinogen (category 2B).

In conclusion, it would be prudent for the medical community to pay closer attention to the possibility of ochratoxin toxicity in patients with symptoms of renal pathology. Although the role of ochratoxin A in human disease is still speculative, its acute nephrotoxicity, immunosuppressive actions, and teratogenic effects in animal models, coupled with its ability to be carried through the food chain, merit concern.

Patulin

Patulin, 4-hydroxy-4H-furo [3, 2c] pyran-2(6H)-one, is produced by many different molds but was first isolated as an antimicrobial active principle during the 1940s from Penicillium patulum (later called Penicillium urticae, now Penicillium griseofulvum). The same metabolite was also isolated from other species and given the names clavacin, claviformin, expansin, mycoin c, and penicidin. A number of early studies were directed towards harnessing its antibiotic activity. For example, it was tested as both a nose and throat spray for treating the common cold and as an ointment for treating fungal skin infections. However, during the 1950s and 1960s, it became apparent that, in addition to its antibacterial, antiviral, and antiprotozoal activity, patulin was toxic to both plants and animals, precluding its clinical use as an antibiotic. During the 1960s, patulin was reclassified as a mycotoxin.

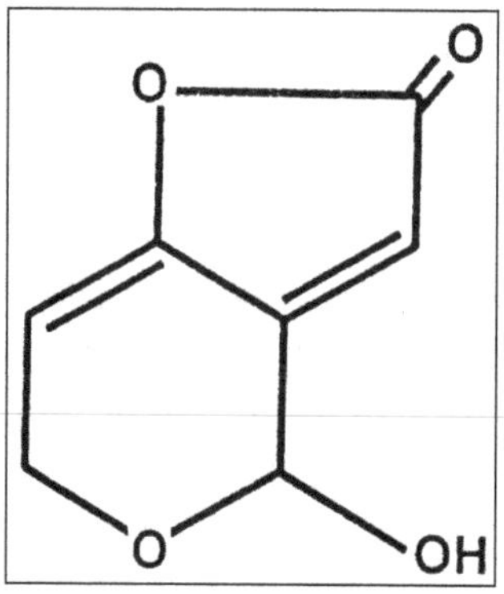

Patulin.

Nowadays, Penicillium expansum, the blue mold that causes soft rot of apples, pears, cherries, and other fruits, is recognized as one of the most common offenders in patulin contamination. Patulin is regularly found in unfermented apple juice, although it does not survive the fermentation into cider products. Patulin is toxic at high concentration in laboratory settings, but evidence for natural poisoning is indirect and inconclusive. Nevertheless, the Joint Food and Agriculture Organization-World Health Organization Expert Committee on Food Additives has established a provisional maximum tolerable daily intake for patulin of 0.4 mg/kg of body weight per day.

Patulin has also played an important role in the study of the classical biochemistry of polyketide biosynthesis. The first cell extract for a fungal polyketide synthase involved studies on a 6-methylsalicylic acid synthase from the species then called Penicillium urticae (now Penicillium griseofulvum).

Trichothecenes

The trichothecenes constitute a family of more than sixty sesquiterpenoid metabolites produced by a number of fungal genera, including Fusarium, Myrothecium, Phomopsis, Stachybotrys, Trichoderma, Trichothecium, and others. The term trichothecene is derived from trichothecin, which was the one of the first members of the family identified. All trichothecenes contain a common 12, 13-epoxytrichothene skeleton and an olefinic bond with various side chain substitutions. They are commonly found as food and feed contaminants, and consumption of these mycotoxins can result in alimentary hemorrhage and vomiting; direct contact causes dermatitis.

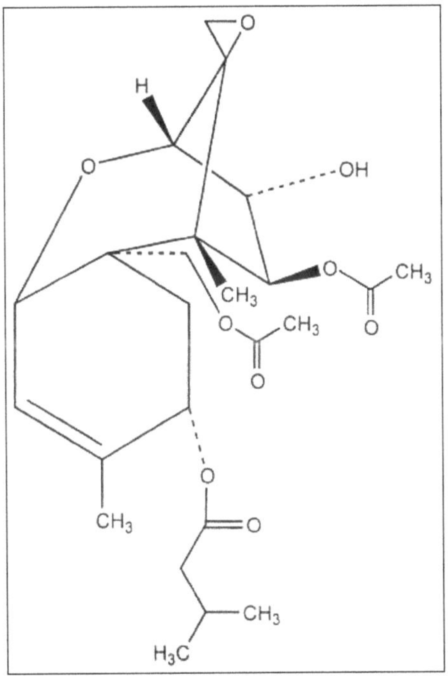

T-2 toxin.

Trichothecenes are classified as macrocylic or nonmacrocyclic, depending on the presence of a macrocylic ester or an ester-ether bridge between C-4 and C-15. The nonmacrocylic trichothecenes in turn can be subclassified into two groups: type A, which have a hydrogen or ester type side chain at the C-8 position, and include T-2 toxin, neosolaniol, and diacetoxyscirpenol, while the type B group contain a ketone and include fusarenon-x, nivalenol, and deoxynivalenol. Fusarium is the major genus implicated in producing the nonmacrocylic trichothecenes. Many members of this genus are significant plant pathogens; their convoluted taxonomy has already been mentioned.

The trichothecenes are extremely potent inhibitors of eukaryotic protein synthesis; different trichothecenes interfere with initiation, elongation, and termination stages. Trichodermin was the first trichothecene shown to inhibit peptidyl transferase activity. Subsequently, it would appear that while all trichothecenes inhibit peptidyl transferase by binding to the same ribosome-binding site, they exert different effects which can be correlated with different functional groups. The 12,13-epoxide group is essential for inhibition of protein synthesis; reduction of the 9,10 double bond reduces toxicity.

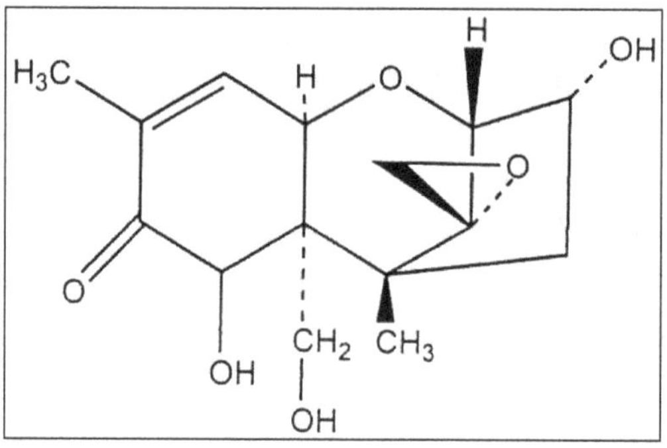

Deoxynivalenol.

There is a long history of moldy grain "intoxications" in Japan, where disease in both human beings and farm animals has been attributed to Fusarium mycotoxicoses. Fusarium graminearum (teleomorph Gibberella zeae), regularly found on barley, oats, rye, and wheat, is considered the most important plant pathogen in Japan and is believed to be the cause of red mold disease (Akakabi toxicosis). As with all mycotoxins, depending on weather conditions, the growth of trichothecene-producing fungi and subsequent production of toxins vary considerably from year to year and from place to place.

Diacetoxyscirpenol, deoxynivalenol, and T-2 are the best studied of the trichothecenes produced by Fusarium species. Deoxynivalenol is one of the most common mycotoxins found in grains. When ingested in high doses by agricultural animals, it causes nausea, vomiting, and diarrhea; at lower doses, pigs and other farm animals exhibit weight loss and food refusal. For this reason, deoxynivalenol is sometimes called vomitoxin or food refusal factor. Although less toxic than many other major trichothecenes, it is the most prevalent and is commonly found in barley, corn, rye, safflower seeds, wheat, and mixed feeds.

The symptoms produced by various trichothecenes include effects on almost every major system of the vertebrate body; many of these effects are due to secondary processes that are initiated by often poorly understood metabolic mechanisms related to the inhibition of protein synthesis.

The macrocyclic trichothecenes are produced largely by Myrothecium, Stachybotrys, and Trichotheciumspecies. Glutinosin, a mixture of the macrocyclic trichothecenes verrucarin A and B, was originally identified as an antimicrobial agent. Recently, the trichothecenes produced by Stachybotrys atra(Stachybotrys chartarum) have received the most attention. They include satratoxins, roridins, verrucarins, and atranones.

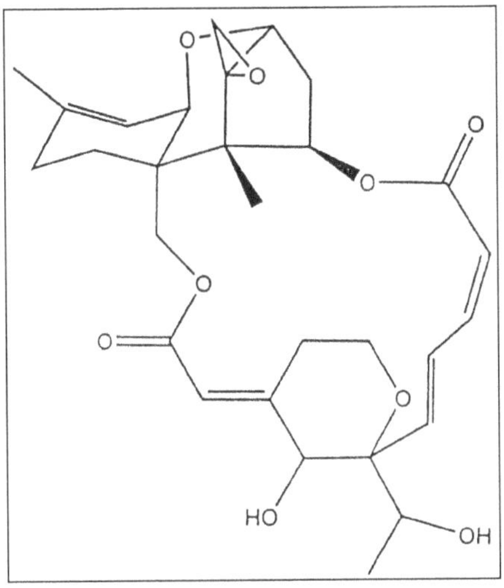

Satratoxin H.

Stachybotryotoxicosis was first described as an equine disease of high mortality associated with moldy straw and hay. Much of the early work on Stachybotrys toxins was published in the Russian literature, focused on horses, and has been well summarized by Forgacs. Until recently, human stachybotryotoxicosis was considered a rare occupational disease limited largely to farm workers who handle moldy hay. However, it has become apparent that Stachybotrys grows well on all sorts of wet building materials with high cellulose content, for example, water-damaged gypsum board, ceiling tiles, wood fiber boards, and even dust-lined air conditioning ducts. The presence of Stachybotryshas been associated with pulmonary bleeding in infants. Although toxic mold was found in the homes of the children with pulmonary bleeding, a cause-and-effect relationship has been difficult to prove. Nevertheless, this incident and a few other highly publicized mold cases seem to have precipitated a series of multimillion-dollar law suits against contractors, real estate developers, and insurance companies as well as a great deal of public apprehension about “toxic molds.”

It was suggested that the 10th plague visited on Egypt was a Stachybotrys infestation, but since the plague infected grain, not hay, it is more likely that Aspergillus or Penicillium was the culprit.

Many of the genes involved in trichothecene biosynthesis in Fusarium sporotrichioides and in Myrothecium roridum have been cloned and sequenced, and there is also considerable progress in understanding the regulation of the trichothecene gene cluster. Homologues of these trichothecene pathway genes have also been reported for Fusarium graminearum.

Zearalenone

Zearalenone (6-[10-hydroxy-6-oxo-trans-1-undecenyl]-B-resorcyclic acid lactone), a secondary metabolite from Fusarium graminearum (teleomorph Gibberella zeae) was given the trivial name zearalenone as a combination of G. zeae, resorcylic acid lactone, -ene (for the presence of the C-1′ to C-2 double bond), and -one, for the C-6′ ketone. Almost simultaneously, a second group isolated, crystallized, and studied the metabolic properties of the same compound and named it F-2. The structure of zearalenone is given in Perhaps because the original work on these fungal macrolides coincided with the discovery of aflatoxins, the word toxin is almost certainly a misnomer because zearalenone, while biologically potent, is hardly toxic; rather, it sufficiently resembles 17β-estradiol, the principal hormone produced by the human ovary, to allow it to bind to estrogen receptors in mammalian target cells Zearalenone is better classified as a nonsteroidal estrogen or mycoestrogen. Sometimes it is called a phytoestrogen.

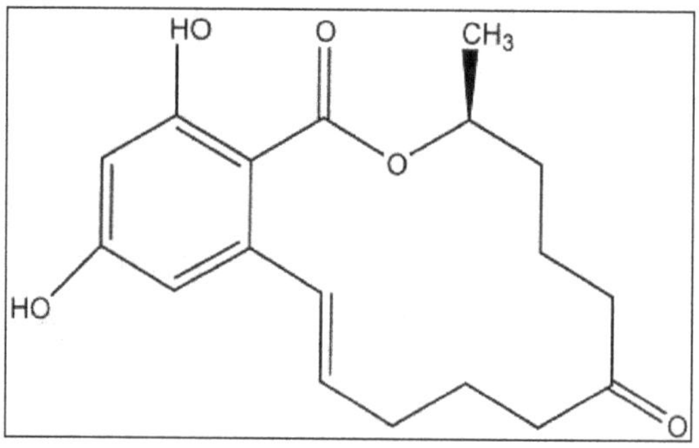

Zearalenone.

The zearalenones are biosynthesized through a polyketide pathway by Fusarium graminearum, Fusarium culmorum, Fusarium equiseti, and Fusarium crookwellense. All these species are regular contaminants of cereal crops worldwide. An association between moldy grain consumption and hyperestrogenism in swine has been observed since the 1920s; modern work shows that dietary concentrations of zearalenone as low as 1.0 ppm may lead to hyperestrogenic syndromes in pigs; higher concentrations can lead to disrupted conception, abortion, and other problems. Reproductive problems have also been observed in cattle and sheep.

The reduced form of zearalenone, zearalenol, has increased estrogenic activity. A synthetic commercial formulation called zeranol (Ralgro) has been marketed successfully for use as an anabolic agent for both sheep and cattle. In 1989, zeranol was banned by the European Union, but it is still used in other parts of the world. Zearalenone has also been used to treat postmenopausal symptoms in women, and both zearelanol and zearalenone have been patented as oral contraceptives. It has been claimed that the high frequency of early menarche in Puerto Rico might be due to zearalenone and related compounds in the human diet; however, studies by the Food and Drug Administration do not support this hypothesis.

Recently, endocrine (hormone) disrupters have received a lot of public attention and are widely believed to reduce male fertility in human and wildlife populations but it is not clear how much the zearalenones contribute to the total environmental load of xenoestrogens. Sometimes, hormone disrupters are labeled environmental toxicants, further muddying the distinction between a compound that can cause death (toxin) and a compound that has other pharmacological activities.

In summary, the zearalenone family of metabolites illustrates some of the limitations of scientific language. The biological potency of these compounds is high, but the actual toxicity is low. The 50% lethal dose in female rats is greater than 10,000 mg/kg; in female guinea pigs it is 5,000 mg/kg, while as little as one µg/kg may create a detectable uterogenic response in female swine. Thus, mycoestrogen is a more appropriate rubric than mycotoxin. Moreover, like the ergot alkaloids, in certain formulations some analogues of these macrolides can be called drugs.

Extensive reviews of Canadian and Scandinavian epidemiological data have concluded that the risk to human populations is minimal. The recommended safe human intake of zearalenone is estimated to be 0.05 µg/kg of body weight per day. Zearalenone levels in foodstuffs are not yet regulated anywhere. Nevertheless, because of its genuine biological potency and regular dietary cooccurrence with true mycotoxins such as the fumonisins and trichothecenes, it would not be prudent to ignore the potential of zearalenone to cause adverse health effects.

Other Mycotoxins and Purported Mycotoxicoses

Penicillium roqueforti and Penicillium camemberti (=Penicillium caseicola), species used to manufacture mold-ripened cheeses, produce a number of toxic metabolites, including penicillin acid, roquefortine, isoflumigaclavines A and B, PR toxin, and cyclopiazonic acid.

Several mycotoxins induce tremors as a neurological response in farm animals; most of these fungal tremorgens contain a modified indole moiety and are produced by certain species of Aspergillus, Penicillium, and Claviceps. The tremorgenic mycotoxins include the penitrems, janthitrems, lolitrems, aflatrem, paxilline, paspaline, paspalicine, paspalinine, and paspalitrem A and B. Penicillium crustosum produces penitrem A, a compound implicated in several cases of canine intoxication and one case of human tremor, vomiting, and bloody diarrhea.

Originally isolated from Penicillium cyclopium (now Penicillium aurantiogriseum), cyclopiazonic acid is an indole tetramic acid. This mycotoxin is a specific inhibitor of calcium-dependent ATPase and induces alterations in ion transport across cell membranes. It is produced by many other species of Penicillium as well as several species of Aspergillus, including Aspergillus flavus. Cyclopiazonic acid was isolated from a sample of the ground nut meal that had been implicated in the original turkey X disease and may have contributed to the severity of that early aflatoxicosis. Furthermore, consumption of a kodo millet that was heavily contaminated with molds and contained detectable levels of cyclopiazonic acid produced kuduo poisoning, characterized by giddiness and nausea. Finally, some strains of Penicillium camembertii involved in the production of gourmet cheese produce cyclopiazonic acid.

The yellow rice toxins (citrinin, citreoviridin, luteoskyrin, rugulosin, rubroskyrin, and related compounds) are believed to have exacerbated Shoshin-kakke, a particularly malignant form of beriberi seen in Japan in the early 20th century.

A number of rare and obscure diseases have been hypothesized to be possible mycotoxicoses, often on extremely meager evidence. These include Kashin-Beck disease in Russia, mselini joint disease and onylalai in Africa, endemic familial arthritis of Malnad in India, frontoethmoidal encephalomeningocele in Myanmar, sago hemolysis in Papua New Guinea, and deteriorated sugar cane poisoning in China. Descriptions of these conditions and their purported association with exposure to toxigenic fungi have been admirably summarized by Beardall and Miller.

ETHNOMYCOLOGY

Ethnomycology is the study of the historical uses and sociological impact of fungi and can be considered a subfield of ethnobotany or ethnobiology. Although in theory the term includes fungi used for such purposes as tinder, medicine (medicinal mushrooms) and food (including yeast), it is often used in the context of the study of psychoactive mushrooms such as psilocybin mushrooms, the Amanita muscaria mushroom, and the ergot fungus.

Amanita muscaria has a long and varied history of psychoactive use.

American banker Robert Gordon Wasson pioneered interest in this field of study in the late 1950s, when he and his wife became the first Westerners on record allowed to participate in a mushroom velada, held by the Mazatec curandera María Sabina. The biologist Richard Evans Schultes is also considered an ethnomycological pioneer. Later researchers in the field include Terence McKenna, Albert Hofmann, Ralph Metzner, Carl Ruck, Blaise Daniel Staples, Giorgio Samorini, Keewaydinoquay Peschel, John Marco Allegro, Clark Heinrich, Jonathan Ott, Paul Stamets, and Juan Camilo Rodríguez Martínez.

Besides mycological determination in the field, ethnomycology depends to a large extent on anthropology and philology. One of the major debates among ethnomycologists is Wasson's theory that the Soma mentioned in the Rigveda of the Indo-Aryans was the Amanita muscaria mushroom. Following his example similar attempts have been made to identify psychoactive mushroom usage in many other (mostly) ancient cultures, with varying degrees of credibility. Another much written about topic is the content of the Kykeon, the sacrament used during the Eleusinian mysteries in ancient Greece between approximately 1500 BCE and 396 CE.

While Wasson views historical mushroom use primarily as a facilitator for the shamanic or spiritual experiences core to these rites and traditions, McKenna takes this further, positing that the ingestion of psilocybin was perhaps primary in the formation of language and culture and identifying psychedelic mushrooms as the original "Tree of Knowledge". There is indeed some research supporting the theory that psilocybin ingestion temporarily increases neurochemical activity in the language centers of the brain, indicating a need for more research into the uses of psychoactive plants and fungi in human history.

The 1990s saw a surge in the recreational use of psilocybin mushrooms due to a combination of a psychedelic revival in the rave culture, improved and simplified cultivation techniques, and the distribution of both the mushrooms themselves and information about them via the Internet. This "mushrooming of mushroom use" has also caused an increased popularization of ethnomycology itself as there are many websites and Internet forums where mushroom references in Christmas and fairy tale symbolism are discussed. It remains open to interpretation what effect this popularization has on ethnomycology in the academic world, where the lack of verifiable evidence has kept its theories with their often far-reaching implications shrouded in controversy.

FORENSIC MYCOLOGY

The use of mycological evidence in criminal investigations and its testing in the courts is called Forensic mycology. It has until recent years largely been restricted to cases involving poisonous and psychotropic species.

Trace Evidence

Like other palynomorphs, fungal spores and other remains may be picked up by any object contacting them and are subject to similar taphonomic considerations. Generally, any palyniferous surface will yield palynomorphs, but the main sources in criminal investigations are soils, sediments, vegetation, and plant litter. Unlike plants, fungi (including lichen fungi) can also grow on, for example, stone, brick, tiles, paving stones, wooden objects, leather, plastics, rubber, and textiles. Their spores may thus provide trace evidence in situations where other palynomorphs are scarce or absent. Even fragments of lichens, or fragments of mouldy objects, can become detached and caught up in items that are involved in criminal investigation.

To find fungal species during their field surveys, mycologists generally rely on their eyesight, hand lens, and experience of which habitats might prove fruitful. We have, however, found that the methods of sampling vegetation, soils, and forensic exhibits for palynomorphs can yield evidence of species of which are extraordinarily rare. For example, we have encountered the distinctive spores of Caryospora callicarpa in preparations made in connection with forensic cases in the UK, although no specimen of the fungus itself has been collected in the country since 1865. This fungus must still be considered rare since it was present in only seven samples out of 1100 examined over a 2-year period, but it is not extinct as might be supposed from field finds. Occurrences of such rare speciesmake them especially valuable as trace evidence.

In an investigation of the murder of a young woman, her body was dumped in a bed of stinging nettles. Nettle (Urtica dioica) can support at least 92 fungi, of which about 17 are known only from this plant. Spores of two fungi common on dead stinging nettles (e.g. Periconia sp., Torula herbarum) were found in palynological preparations from the crime scene and also in the suspect's car. These supported ecological and palynological evidence demonstrating a link between the suspect and the place where the corpse had been deposited.

In a drug-related shooting, a gunman hid himself up against the trunk of an oak tree that was growing in a cypress hedge along a quiet lane in Romford, Essex and killed an associate who had arranged to meet him in the lane. The cypress was growing sub-optimally because of intense shading by the oak, and was infected with a pathogenic fungus (Pestalotiopsis funerea). The gunman had been standing in deep leaf and twiggy litter, and his body and feet could not avoid contacting the cypress branches, oak trunk, and the litter. The pollen assemblage from the crime scene was very similar to that found associated with items seized from the suspect and his associates. The spores of Pestalotiopsis funerea formed part of that palynomorph assemblage; they were abundant in the leaf litter and vegetation samples, and were found in a vehicle known to have been used by the gunmen and associates. In addition, spores of what most probably was an as yet undescribed Endophragmiella species were also found in the leaf litter and the get-away car. This was a case where, although the palynological profiles of the comparator samples and exhibits were similar, the fungal spores provided additional resolution that added powerful pertinent information for the court.

Another criminal case, where a relatively rare fungus provided great resolution to the palynological profile, was in the first successful prosecution for badger-digging for the Royal Society for the Prevention of Cruelty to Animals (RSPCA). Again, the palynological profile from soil samples obtained from the badger sett was very similar to that retrieved from a spade and shovel that had been seized from a suspect. Badgers had made their sett in a pasture very close to a somewhat neglected hedge in the Staffordshire countryside. Two large oak trees were growing in the boundary hedge, each about 100 m from the sett. It was very surprising to find spores of the white-veined truffle (Choiromyces meandriformis) in the soil of the sett and in the soil adhering to the suspect's spade. The truffle is the sporophore of the fungus, and this forms beneath the soil surface in association with the roots of a variety of deciduous trees, including oak. It is well known that pigs, dogs, and humans enjoy truffles, but here was a case where badgers must have been digging for them in the oak's root system and taking them back to their sett. This rare spore helped link that spade and the badger sett.

Although occasional rare fungal spores enhance the resolution of the palynological data, when whole fungal and large pollen and plant spore assemblages are combined, evidence of contact becomes very powerful. Each fungal species has particular nutritional and habitat requirements, so assemblages of fungi can indicate very specific situations. In Lincolnshire in 2008, four animal rights activists were arrested on suspicion of being involved in an attempted break-in at a farm that bred laboratory rabbits. Footwear from each of the suspects yielded palynological profiles that were similar to those in the comparator samples from the farm. It was of interest that exotic plant species were represented, and it was interpreted that their pollen had been derived from the imported rabbit food. In addition, the comparator samples and footwear yielded 21 fungal taxa which were identified to genus or species. Of these, were associated with the suspects' footwear; and three of these have few records in the FRDBI; they must, therefore, be considered rare. These were Brachysporiella setosa, Caryospora callicarpa, and Melanospora zamiae. The latter species is mostly found in warmer regions of the world. Like some of the pollen taxa, it is possible that at least some of the fungal species were derived from the imported rabbit food. In any event, the complex, combined assemblages contributed to a specific trace signature from the farm that was also found on the footwear of the suspects. With so many trace markers being involved, and considering the rarity of some of these, the likelihood of this specific assemblage being obtained from any other location was remote.

While the precision of pollen data is enhanced by fungal spores, and they provide a heightened level of correlation between comparator samples and objects retrieved from suspects, they are generally found in low concentration in palynological preparations. This means that large numbers of palynomorphs need to be counted in order to even find the fungi. In our experience, it is often the rare spores found in palynological preparations that enhance the value of this class of trace evidence. This is demonstrated by a rape case investigated by Wiltshire Constabulary in 2009. The victim claimed she had been raped on the ground under some trees, while the suspect insisted they had had consensual sex on grass in a park about 200 m away. Nineteen fungi characteristic of dead leaves and twigs of woody plants were found on the clothing and footwear of victim and suspect. Of these, 16 were represented in the comparator samples from the area under the trees, and only four from the park. When combined with the other palynological results, the fungal assemblage clearly showed that, despite their close proximity, the two sites could be easily differentiated. The results indicated that the woman was telling the truth about where the incident took place, but such convincing findings needed a palynomorph count of over 38,000. The outcome was rewarding for investigating officers since, when confronted with the palynological evidence, the suspect confessed.

In the present century, it might be thought that DNA profiles of environmental samples that include fungi also had potential as trace evidence. The reality is that the methods that have been attempted are not yet sufficiently robust and reproducible for acceptance in a court of law.

Fungi	Suspect	Victim	Wood	Park
Clasterosporium flexum	+	+	+	
Pseudovalsella-like	+	+	+	
Pestalotiopsis funerea	+	+	+	+
Brachysporium britannicum	+	+	+	
Camposporium cambrense		+	+	
Diplocladiella scalaroides	+		+	
Glomus-type	+		+	
Didymosphaeria sp.		+	+	+
Melanospora sp.		+	+	
Bactrodesmium obovatum	+		+	
Dictyosporium toruloides	+		+	
Bactrodesmium betulicola	+		+	
Periconia byssoides	+	+	+	+
Epicoccum nigrum	+	+	+	+
Endophragmiella fagicola	+		+	
Asterosporium hoffmannii	+			
Cymadothea trifolii				+
cf. Diporotheca sp.	+			
Phaeotrichosphaeria brittanica	+			
Niesslia exosporioides	+			
Ball of straw coloured cells		+	+	
Palynomorphs counted	3236	1680	3294	30,129

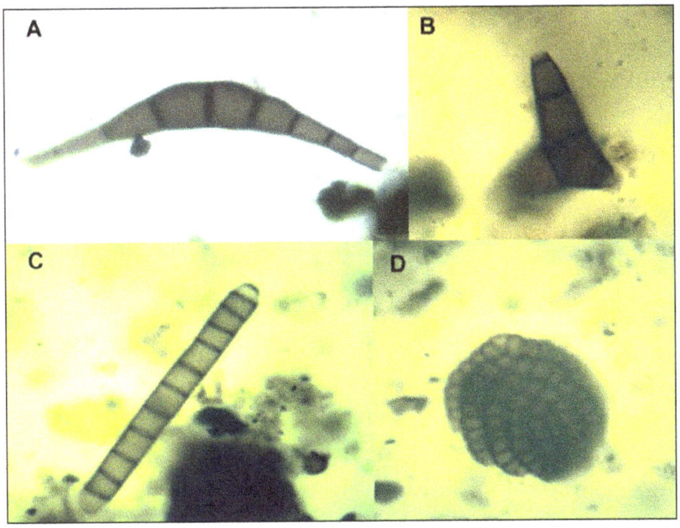

A selection of fungal spores found in palynological preparations in a rape case investigated for Wiltshire Constabulary in 2009: (A) Clasterosporium flexum; (B) Diplocladiella scalarioides; (C) Camposporium cambrense; and (D) Dictyosporium toruloides.

Various molecular methods of characterising fungal communities in soil, in which species are not identified but "fingerprint"-like patterns are produced, have been tried. But, at the present time, none is at a stagewhere it could be commended for either ecological or forensic investigation. Fungi in soils are particularly diverse; 50– 100 different species can be expected to be isolated into culture from a single soil sample, and there can be as much as 3 kmof fungal hyphae in 1 g of dry soil. This diversity makes the development of standard molecular, or other, approaches extremely difficult. Currently, precise molecular characterisation of any soil is impractical for a variety of reasons, including the following: (1) A single molecular approach is unlikely to be universally applicable. (2) The commonly used terminal restriction fragment length polymorphism (TRFLP) approach is unreliable for due to several problems, not least being peaks due tomultiple fragments as a result of PCR enzyme inefficiency, and also whether conserved or more variable genes are targeted. (3) There can be huge differences in profiles obtained fromsample sites in close proximity;one studyhas shownthat soils associatedwith the same tree species growing 20m apart were very different. (4) The amount of soil from which total DNA is extracted for analysis in molecular ecology studies is generally 250–500 g – amounts hardly ever encountered on forensic exhibits; sometimesmere traces are retrieved fromexhibits. (5) Relatively few fungi are represented by correctly named species in public DNA databases; this makes species-level identifications problematic. There are, however, possibilities for the detection of particular species of fungi in environmental samples using specially designed probes, but this approach requires prior knowledge of what is present.

We have not traced work aimed at determining the extent of reproducibility in molecular profiles obtained from a single site, but we suspect that this would be low due to the abundance and wide variety of fungal propagules to be expected in a single sample.

Time Since Death (Post-mortem Interval)

Although healthy humans can have fungal infections, the fungi involved are usually specialised species tolerant of high body temperatures and human immune systems. They range from dermatophytic fungi, which cause ringworm on the surface of the skin, to invasive infections, such as candidiasis (thrush), and deepseated infections in lungs (e.g. aspergillosis) and other tissues (e.g. mycetomas, mycoses). In immunocompromised individuals, however, a variety of less specialised fungi can occasionally occur opportunistically within human tissues. In our experience, the fungi frequently found growing on and in corpses are those that are not normally able to colonise living tissue. Surprisingly, there is almost no information on the role of particular fungi in the decomposition of human corpses. However, remains interred directly in soil frequently have been mentioned as showing signs of moist decomposition, with skin slippage and fungus development. Janaway et al. commented that soil fungi might be involved on the body surface "after the major phase of decomposition", and mention that "moulds begin to appear on the surface of the body" in the first week. No source for these statements was given although our own work confirms his observations. Janaway et al. also note that an unidentified Candida species had been isolated from an early stage of decomposition (again with no source indicated), and that fungi can be found growing in soil infused with decomposition products – citing Sagara et al., and also a paper mentioning two Penicillium species.

The only experimental study appears to be that of Parkinsonet al. who endeavoured to compare the changes in fungal communities in soil in response to human cadaver deposition on the surface. This study was carried out at the Forensic Anthropology Center of the University of Tennessee, and used molecular approaches; the control samples appeared "to cluster a little more closely than the cadaver samples". No clear trend in succession was evident, and no attempt was made to identify the fungi detected.

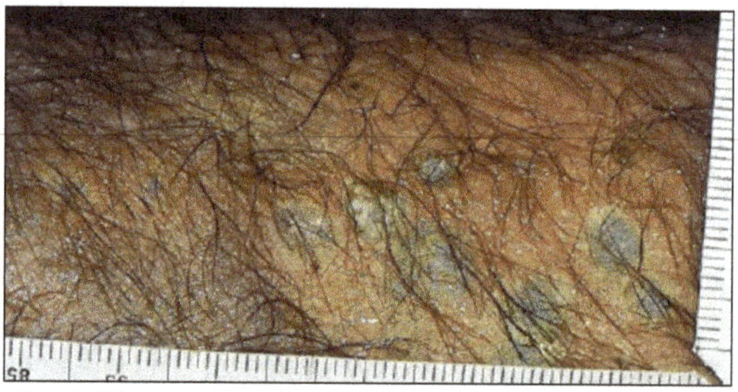

Colonies of Penicillium griseofulvum growing on abdominal skin of a cadaver.

The fungi that have proved of most interest in estimating post-mortem interval are neither specialised medically important fungi, nor ones restricted to dead human tissues, but rather are decomposer or spoilage fungi that are able to directly colonise the surfaces of corpses and body parts after death. It is not surprising that fungi have occasionally been noted by pathologists on the surface of human cadavers. One key reference text for forensic pathologists reports extensive growths on a boy after 6 weeks, while another refers to development after several months on an embalmed cadaver. However, the fungi involved have not generally been identified even to genus, nor considered as a forensic tool.

The first researchers to appreciate that fungal growth on corpses had a role in the determination of time of death were van de Voorde and van Dijck. These authors made isolations from fungal growths on an eyelid and on inguinal skin of a murdered baroness found in a room in Belgium. They incubated the fungal isolates at the same temperature as that in which the body had been found (a constant 12 8C controlled by a thermostat in the room), and measured the colony sizes daily. They estimated that the woman had died at least 18 days before her body was discovered, and that agreed exactly with a subsequent admission by the murderer. They considered that superficial fungi could help in determining the time of death, when this was 10–20 days earlier, provided those temperature data were available. The fungi involved in this case were named as: Cladosporium sp., Fusarium sp., Geotrichum candidum, Hormodendron sp., Mortierella sp., and Penicillium chysogenum (as P. notatum).

Unaware of the Belgian case, Ishii et al. reported on fungal growths found on the surface of a mummified body found in an abandoned house, and also on skeletal remains discovered in a forest, that they suggested "may reveal local habitats". They identified these as Aspergillus chevalieri (as Eurotium chevalieri), A. repens (syn. E. repens) that was the most abundant, A. rubrum (as E. rubrum), and Gliocladium sp. (probably actually a species of Clonostachys). These authors recognised that more cases were needed to establish the approach as a forensic tool, but did not attempt to estimate PMI in these cases. They subsequently reported on white growths dotted on the face of a male corpse found in a kneeling position and partly in water at the bottom of an open well in Japan. The humidity was nearly 100%, and the temperature constant at 12–13 8C; the fungi were isolated into culture and identified as Aspergillus terreus and Penicillium sp. The deceased had been dead for 10 days, and although no growth measurements of the colonies were made, they suggested that the fungal data were consistent with the time as the fungi "generally colonise in 3–7 days" – a generalization based on the time-lag in attacking foodstuffs. These workers independently suggested that fungi could provide a useful means of estimating PMI where forensic entomological data were not available.

Menezes et al. considered that Ishii et al.'s claim of fungal growths being of value in estimating PMI's would have been more convincing if they had investigated the growths empirically. Their response reaffirmed their view of the potential, acknowledging that there were unsolved problems regarding growth rates and the actual fungi. In a rejoinder, Menezes et al. stressed the need for more forensic cases to be documented from different climatic conditions, and for experimental work on a substitute subject, especially in relation to the pattern and rate of growth of fungal species on cadavers.

When the body of a man was found by a tube line in Ruislip, north-west London, the medical examiner considered it had been there not more than 48 h. There was no evidence of any scavenger or insect activity. However, botanical evidence, and a large circular fungal colony under the chin, indicated that the body had lain in situ for between 3 and 5 weeks. It later transpired that the actual time was 4 weeks 3 days. Decomposition had been delayed because the weather had been exceptionally cold, and the low temperatures had inhibited colonisation by flies. Security fencing had excluded scavenging animals and the body had remained intact.

In the case of the serial killing of prostitutes around Ipswich, Suffolk at the end of 2006, two of the bodies were recovered from water. These were ensheathed in fungal growth that had trapped silt particles that had become graded as they accumulated in the hyphal10 weft. Several fungi were involved, including species of Fusarium, Geotrichum, Mucor, and Pythium. The first body that was retrieved was completely ensheathed. With the second body, the sheath was present though it was patchy and had collapsed due to the much greater length of the hyphae than on the other body. Where there had been skin slippage and exposure of the underlying dermis, there had been fresh fungal colonisation on some of the exposed areas. The fungi were intermixed and not identified to species. Even though information on the growth of these fungi in cold water was not available, it was obvious that the body of this victim had been in the water longer than that of the other. An estimate of at least 5 weeks submersion was suggested by the condition of the fungal mycelium and other features of this corpse. Using the same criteria, it was suggested that the woman who had been found first had been in the water for about 2 weeks. These estimates agreed with the length of time the victims had been reported to be missing.

Fungal colonies on or associated with, human cadavers can give indications of time since death as there is information on growth rates of many moulds. But the reliability of any estimates will depend on the accuracy of the identification of the fungus, the storage methods for the body, and the availability of data on the temperature and humidity at the site. As yet there are few precise data on actual rates of growth on dead human tissues, especially under different conditions of temperature and humidity. This means that it is necessary to undertake experiments mirroring, as near as is possible, the environmental parameters with which the corpse was associated, using the actual strains found. We are currently involved in such experimental work.

In a case undertaken for Tayside Police in 2009, a man had been repeatedly stabbed and died in a closed flat so that flies were excluded. Fungal colonies had developed on parts of the carpet and on a sofa that had become soaked with body fluids. After the discovery and removal of the body, data loggers were placed in the room and consistently gave relative humidity readings of 30–34%. Colonies were measured in situ and subcultures made onto artificial media. Samples of the carpet were then removed and kept dry for observation; no subsequent growth occurred within 5 days. This result was expected since most mould fungi require at least 95% relative humidity for growth. When the carpet samples were re-wetted with bovine blood and kept at a high humidity, there was a spurt of new fungal growth. A comparison of the sizes of the new colonies, both in culture and on the carpet, with those at the crime scene, suggested that the death had occurred about 5 days prior to discovery of the body; this was consistent with a subsequent admission of guilt. The three principal fungi involved in this investigation were Mucor plumbeus, Penicillium brevi-compactum, and P. citrinum.

Carpet saturated with body fluids and then dried, showing fungal colonies.
The large, dark greyish colonies are of Mucor plumbeus.

Time of Deposition

Exposed bones can be colonised by mould and other fungi under appropriate ambient conditions. Although we could find no modern published information on fungi growing on human bones, conspicuous fungal colonies were found on the scavenged skeletal remains of a woman whose remains had been found in a Sussex woodland in 2008; 13 fungal isolates were obtained from the leg and pelvic bones. It must be noted that there was still a very thin and patchy film of soft tissue on some of the bones and some were also rather greasy. These fungal isolates are the subject of current research.If bone is exposed for many years in well-lit situations, lichen colonies of, in particular, Caloplaca and Lecanora species can develop. Colony diameter can give a good indication of minimum periods of exposure. There are also ancient reports of lichens growing on human skulls. In particular, the skull lichen (Parmelia saxatilis or some related species) was supposed to be "worth its weight in gold" as a cure for epilepsy. In 1640 Parkinson stated that it "groweth upon the bare scalps of men and women that have lyne long", and refers to reports that for greatest efficacy "it should be taken from the sculls of those who have been hanged or executed for offences". There is a huge literature on the use of lichens in dating surfaces by geomorphologists, glaciologists and, to a lesser extent, archaeologists. However, these mostly use the "map lichens" of the Rhizocarpon geographicum complex and these do not grow on calcareous materials.

In a recent case we found it possible to provide information on the time that a plastic-wrapped body part had lain in situ from botanical and lichen evidence. A broken twig that had the lichen Xanthoria parietina growing on it was on the ground under the body part. When well illuminated, this lichen is yellow-orange with deeper orange sporophores. When growing naturally in the shade, even on the underside of the same twig, the lichen is grey but still has the orange sporophores. However, if kept in the dark and in too-moist conditions, it will turn green. A simple experiment was constructed whereby a sample of the lichen from the same locality was covered by a plastic-covered weight; compared with an adjacent but uncovered control, it changed from yellow to green within 5 days. This indicated that the body part at the crime scene had probably not been in position for more than a week. Some other lichens can also change colour under adverse conditions.

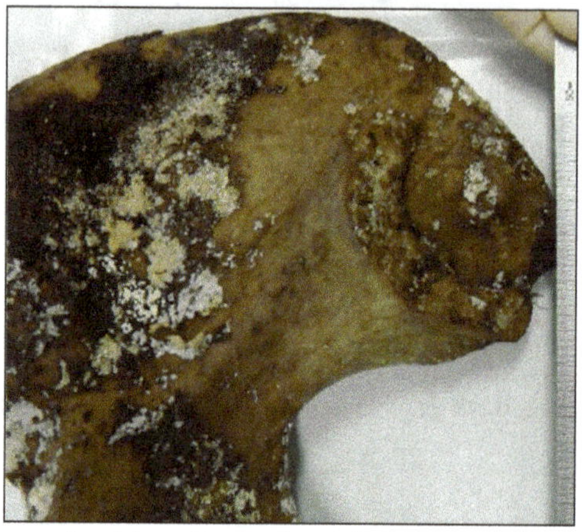

An example of fungal growth on defleshed human bone.

The sporing cycle of fungi can also help in timing events. The scavenged skeleton of a young female was found in a shallow grave close to the A40 west of London in February 2004. It appeared that she had lain there for a long time. However, the remains lay on some shooting bramble stems that indicated she could not have been there for more than 1 year. Also, there were some detached and still-green leaves of bramble (Rubus fruticosus) in the grave fill. The underside of the leaves had black pustules while the leaf tissue associated with them was red-purple. The pustules were caused by a common rust fungus, Phragmidium violaceum that produces orange spores in the spring and early summer (urediniospores), and dark ones (teliospores) in the late summer and autumn.

Many mushrooms are also seasonal, and this is not only a matter of species that form sporophores in the autumn. Some are much more restricted, for example, the scar-let-cupped Sarcoscypha species appear in the early spring on fallen twigs. These fungi have distinctive spores and might provide important trace evidence if trodden on by suspects. Another potentially useful fungus is the mushroom Flammulina velutipes, that only appears after the first frosts of winter.

Fungal colonies grow in a circular manner when on solid substrates, and the growth rates on artificial media are often cited in species descriptions. Those with non-septate hyphae belonging to the Mucorales (pin-moulds) are, in general, much more rapidly growing than those belonging to other groups (e.g. ascomycetes, basidiomycetes) that have repeatedly septate hyphae. In a recent case for Hertfordshire Constabulary, a mutilated murder victim was deposited on his ventral surface on a steep bank above a small, sluggish stream. The absence of blood indicated that the murder and mutilation had occurred elsewhere. Small colonies of nonsporing Mucor hiemalis had developed on the abdomen. Palynological analysis showed that mud that covered the corpse had been derived from the stream and the victim had obviously been dropped in or dragged through it. The colony size and immaturity of the fungus indicated 1–2 days growth, and the pristine condition of the mycelium showed that it had developed in situ after the body had been dragged through the stream. Experience has shown that fungi, such as Mucor species, are unable to colonise and grow on freshly dead skin and it would appear, on average, that this is unable to happen for at least a week after death. This reinforced the hypothesis that the victim had been killed and stored elsewhere and had lain at the deposition site for a short time.

Developing Forensic Mycology

While mycology has been demonstrated to provide useful forensic evidence in a variety of ways and, in some cases, has proved critical in securing convictions, it is currently rarely employed. The principle reason for this is a lack of awareness amongst crime scene investigators, and investigating officers, that fungi have the potential to make significant contributions. In several of the cases in which we have been involved, it was the astuteness of officers in recognising there was fungal growth present on human remains or artefacts that led to our being called in.

A secondary reason is a shortage of appropriately skilled mycologists. Some categories of mycological investigation, particularly those involving the isolation and culture of fungi from remains and materials, can be undertaken by biologists with experience in microbiological methods, but those requiring identification need specialists of wide experience. There is a shortage of professional mycologists able reliably to identify fungi across different systematic groups. For example, in the UK, there has been a reduction in systematic mycology posts in the last 10– 15 years as a result of the restructuring of research institutes and retirements. The number of full-time fungal (including lichen) systematists in the country (excluding curators of collections) is currently a mere eight (of which two are lichenologists); and not one is now employed anywhere in the university sector. This figure of eight compares with 23 in 1997, and the House of Lords Science and Technology Committee has considered the situation "so grave as to be generally recognised as a crisis". The UK is not unique in this respect, and action is needed to redress this problem so as to ensure that adequate support is available not only for forensic science, but also food safety, human health, plant and tree diseases, pharmaceutical development, environmental assessments, and nature conservation. No quick-fix is possible as it takes at least a PhD, plus 10 year's experience, to produce a competent professional systematic mycologist.

The subject is also difficult because of a lack of comprehensive identification manuals. This makes it hard for a non-specialist to make critical identifications. Although there are monographic treatments of some groups, even to know which work to use requires specialist knowledge. Because the number of species is high and many are still unknown, this is a particular problem in fungi, compared with, for example, plants. Furthermore, previously unknown or unrecognised species are discovered every year. Semipopular guides rarely treat all species of a genus, and often do not indicate that there are other similar species. There is not even a pocket-sized identification manual on all hallucinogenic fungi, something that would make the psilocybin and psilocin possession section of the Drugs Act 2005 more practical for police officers to enforce. Those undertaking aspects of forensic mycology that involve numerous identifications, as in the provision of trace evidence, consequently require access to a substantial library and to reference collections of authoritatively named material. The most reliable way to identify fungi is by direct comparison with correctly named reference material.

Identifications using molecular sequence data can be made in certain well-researched groups of fungi, especially those involved in food spoilage, human diseases, or plant pathology. The existing DNA databases, however, have only a small proportion of the known fungi represented in them, and many are represented by only one or a few strains, so there is little representation of within-species variability. Further, the gap between those known, and those that are sequenced, may be growing; only 21% of the species described in 2003–07 are represented in GenBank. Also, and especially worrying, is that several independent studies indicate that around 20% of the fungal sequences deposited in GenBank are based on wrongly identified material. This means that while in some cases DNA data may be able to provide conclusive identifications, in many others either it will not, or it will yield erroneous results. Specialist interpretation is, therefore, essential in using DNA (e.g. BLAST) approaches for fungal identification; this is not something that is yet at a stage where it can be delegated to non-specialists.

References

- Lichenology, science-biology-branches-of-biology: softschools.com, Retrieved 02 June, 2020
- Heinrich, Clark (2002). Magic Mushrooms in Religion and Alchemy (second ed.). Park Street Press. p. 256. ISBN 0-89281-997-9
- Margulis, Lynn; Barreno, EVA (2003). "Looking at Lichens". BioScience. 53 (8): 776. doi:10.1641/0006-3568(2003)053[0776:LAL]2.0.CO;2
- Forensic-mycology-the-use-of-fungi-in-criminal-investigations-45199583: researchgate.net, Retrieved 28 April, 2020

Biological Classification of Fungi

3 CHAPTER

Fungi can also be classified on the basis of their biological characteristics. Biological classification of fungi includes basidiomycota, ascomycota, neocallimastigomycota, blastocladiomycota, glomeromycota, chytridiomycota, microsporidia, etc. This chapter has been carefully written to provide an easy understanding of such biological classifications of fungi.

BASIDIOMYCOTA

Basidiomycota is a major division (or phyla) of the kingdom Fungi, whose members typically are characterized by the presence of a basidium, a microscopic reproductive structure where sexual spores are produced. The presence of basidia ("little pedestal") is one of the main diagnostic features of the Basidiomycota and is the source of the group's name. However, Basidiomycota also includes single-celled forms (yeasts) and asexual species. Basidiomycota commonly are known as club fungi, given the club-shaped basidium. They also are known as basidiomycetes based on some older classifications that recognize Basidiomycetes as a class-level taxon.

About 30,000 described species are placed within Basidiomycota, or about 37 percent of all described species of fungi. Basidiomycota is so varied that unique morphological features of the group, such as basidium, dikaryon, and clamp connections, are not found in all representatives. Well-known Basidiomycota include rusts, smuts, various yeasts, true mushrooms, jelly fungi, false truffles, puffballs, stinkhorns, bracket fungi, boletes, chanterelles, earth stars, bunts, mirror yeasts, and the human pathogenic yeast, Cryptococcus.

Basidiomycota are very important for the ecosystem and for humans. Ecologically, they are vital for decaying dead organic matter, including wood and leaf litter, and thus vital for the carbon cycle. Some also form important symbiotic relationships, such as mycorrhizal associations with the roots of a plant, whereby the fungus receives carbohydrates from the plant's photosynthesis and the plant gains the mycelium's very large surface area to absorb water and mineral nutrients from the soil. Some are cultivated by ants. For humans, some Basidiomycota are a source of food. On the other hand, they also will decompose living wood tissue and thus can damage the wood in homes and other buildings, and the rusts and smuts are parasitic on plants and cause diseases of important agricultural crops, such as wheat. Some cause human and animal diseases.

Overview and Description

Basidomycota are one of the major groups of the fungi kingdom. Fungi are eukaryotic organisms that digest their food externally and absorb the nutrient molecules into its cells. Unlike plants, they do not make their own food through photosynthesis and unlike animals they are not mobile and absorb nutrition externally rather than internally.

Fungi may be single-celled or multicellular. Multicellular fungi are composed of networks of long hollow tubes called hyphae. The hyphae often gather in dense networks known as mycelium. The mycelium is the vegetative part of a fungus, and consists of the mass branching, thread-like hyphae. The mycelium grows through the medium on which the fungus feeds. Because most fungi are embedded in the medium in which they grow, most often soil, they are often not visible.

Although fungi lack true organs, the mycelia can become organized into more complex reproductive structures called fruiting bodies, or sporocarps, under the right conditions. The sporocarp or fruiting body is a multicellular structure in which spore-producing structures, such as basidia or asci, are borne. The fruiting body is part of the sexual phase of a fungal life cycle, the rest of the life cycle is characterized by its vegetative mycelial growth. The sporocarp of a basidiomycete is known as a basidiocarp, while the fruiting body of an ascomycete is known as an ascocarp. Mushroom is the common name given to the fruiting bodies of many fungal species. Although these typically above-ground structures are the most conspicuous to humans, they make up only a small portion of the entire fungal body.

The major divisions (phyla) of fungi are mainly classified based on their sexual reproductive structures. Currently, four or five main divisions are recognized, one of which is the Basidomycota or club fungi. Another major division is the Ascomycota, or sac fungi. Together, the Basidomycota and Ascomycota are commonly known as "higher fungi" (subkingdom Dikarya).

Description of Basidomycota

Although considered monophyletic, Basidiomycota is very varied, with unicellular and multicellular, sexual and asexual, and terrestrial and aquatic forms. As a result, "it is impossible to identify any morphological characteristics that are both unique to the group and constant in the group".

The production of basidia is the most diagnostic feature. A basidium is a cell in which nuclear fusion and meiosis occur prior to developing the typically haploid basidiospores. However, basidia are limited to the sexual Basidiomycota. Another characteristic feature is a long-lived dikaryon, in which all the cells in the thallus contain two haploid nuclei as a result of a mating event. Clamp connections, a kind of hyphal outgrowth, are another unique feature to Basidiomycota.

Basically, Basidiomycota are filamentous fungi composed of hyphae (except for those forming yeasts), and reproducing sexually via the formation of the specialized club-shaped end cells (basidia) that normally bear external spores (usually four), which are the specialized meiospores known as basidiospores. Asexually reproducing Basidiomycota can be recognized as members of this phylum by gross similarity to others, by the formation of the distinctive anatomical feature known as the clamp connection, by components of the cell wall, and definitively by phylogenetic molecular analysis of DNA sequence data.

Classification

The most recent classification adopted by a coalition of 67 mycologists recognizes three subphyla (Pucciniomycotina, Ustilaginomycotina, Agaricomycotina) and two other class level taxa (Wallemiomycetes, Entorrhizomycetes) outside of these, among the Basidiomycota. As now classified, the subphyla join and also cut across various obsolete taxonomic groups previously commonly used to describe various Basidiomycota.

The Basidiomycota had traditionally been divided into two obsolete classes, the Homobasidiomycetes; and the Heterobasidiomycetes (the Jelly, Rust, and Smut fungi). Previously, the entire Basidiomycota were called Basidiomycetes, an invalid class level name coined in 1959 as a counterpart to the Ascomycetes, when neither of these taxa were recognized as phyla. The terms basidiomycetes and ascomycetes are frequently used loosely to refer to Basidiomycota and Ascomycota. They are often abbreviated to "basidios" and "ascos" as mycological slang.

The Agaricomycotina includes what had previously been called the Hymenomycetes (an obsolete morphological based class of Basidiomycota that formed hymenial layers on their fruitbodies), the Gasteromycetes (another obsolete class that included species mostly lacking hymenia and mostly forming spores in enclosed fruitbodies), as well as most of the jelly fungi.

The Ustilaginomycotina are most (but not all) of the former smut fungi, along with the Exobasidiales.

The Pucciniomycotina includes the rust fungi, the insect parasitic/symbiotic genus Septobasidium, a former group of smut fungi (in the Microbotryomycetes, which includes mirror yeasts), and a mixture of odd, infrequently seen or seldom recognized fungi, often parasitic on plants.

Two classes, Wallemiomycetes and Entorrhizomycetes cannot at present be placed in a subphylum.

Typical Life Cycle

Unlike higher animals and plants, which have readily recognizable male and female counterparts, Basidiomycota (except for the Rust (Pucciniales)) tend to have mutually indistinguishable, compatible haploids, which are usually mycelia being composed of filamentous hyphae.

Typically, haploid Basidiomycota mycelia fuse via plasmogamy and then the compatible nuclei migrate into each other's mycelia and pair up with the resident nuclei. Karyogamy is delayed, so that the compatible nuclei remain in pairs, called a dikaryon. The hyphae are then said to be dikaryotic. Conversely, the haploid mycelia are called monokaryons. Often, the dikaryotic mycelium is more vigorous than the individual monokaryotic mycelia, and proceeds to take over the substrate in which they are growing. The dikaryons can be long-lived, lasting years, decades, or centuries.

The monokaryons are neither male nor female. They have either a bipolar (unifactorial) or a tetrapolar (bifactorial) mating system. This results in the fact that following meiosis, the resulting haploid basidiospores and resultant monokaryons have nuclei that are compatible with 50 percent (if bipolar) or 25 percent (if tetrapolar) of their sister basidiospores (and their resultant monokaryons) because the mating genes must differ for them to be compatible. However, there are many variations of these genes in the population, and therefore, over 90 percent of monokaryons are compatible with each other. It is as if there were multiple sexes.

The maintenance of the dikaryotic status in dikaryons in many Basidiomycota is facilitated by the formation of clamp connections that physically appear to help coordinate and re-establish pairs of compatible nuclei following synchronous mitotic nuclear divisions. Variations are frequent and multiple. In a typical Basidiomycota lifecycle, the long lasting dikaryons periodically (seasonally or occasionally) produce basidia, the specialized usually club-shaped end cells, in which a pair of compatible nuclei fuse (karyogamy) to form a diploid cell. Meiosis follows shortly with the production of 4 haploid nuclei that migrate into four external, usually apical basidiospores.

Variations occur, however. Typically the basidiospores are ballistic, hence they are sometimes also called ballistospores. In most species, the basidiospores disperse and each can start a new haploid mycelium, continuing the life cycle. Basidia are microscopic, but they are often produced on or in multicelled large fructifications called basidiocarps or basidiomes, or fruitbodies), variously called mushrooms, puffballs, and so forth. Ballistic basidiospores are formed on sterigmata, which are tapered spine-like projections on basidia, and are typically curved, like the horns of a bull. In some Basidiomycota, the spores are not ballistic, and the sterigmata may be straight, reduced to stubs, or absent. The basidiospores of these non-ballistosporic basidia may either bud off, or be released via dissolution or disintegration of the basidia.

Schematic of a typical basidiocarp, the dipoid reproductive structure of a basidiomycete, showing fruiting body, hymenium and basidia.

Meiosis takes place in a diploid basidium. Each one of the four haploid nuclei migrates into its own basidiospore. The basidiospores are ballistically discharged and start new haploid mycelia called monokaryons. There are no typical males or females, rather there are compatible thalli with multiple compatibility factors. Plasmogamy between compatible individuals leads to delayed karyogamy leading to establishment of a dikaryon. The dikaryon is long lasting but ultimately gives rise to either fruitbodies with basidia or directly to basidia without fruitbodies. The paired dikaryon in the basidium fuse (i.e karyogamy takes place). The diploid basidium begins the cycle again.

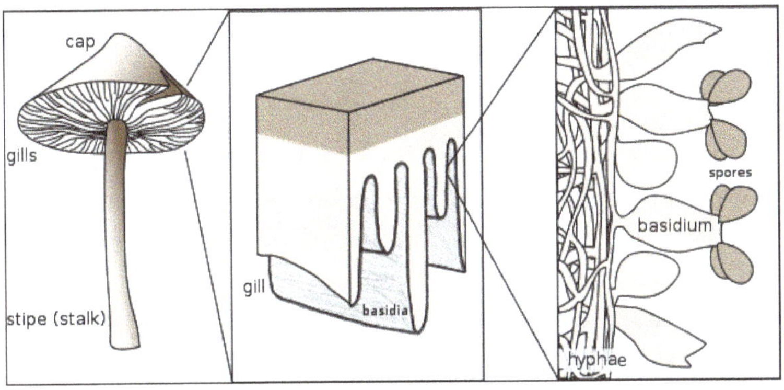

Schematic of a typical basidiocarp, the dipoid reproductive structure of a basidiomycete, showing fruiting body, hymenium and basidia.

Variations in Life Cycles

There are many variations in the Basidiomycota life cycle from the typical cycle. Some Basidiomycota are self compatible and spontaneously form dikaryons without a separate compatible thallus being involved. These fungi are said to be homothallic versus the normal heterothallic species with mating types. Others are secondarily homothallic, in that two compatible nuclei following meiosis migrate into each basidiospore, which is then dispersed as a pre-existing dikaryon. Often such species form only two spores per basidium, but that too varies. Following meiosis, mitotic divisions can occur in the basidium. Multiple numbers of basidiospores can result, including odd numbers via degeneration of nuclei, or pairing up of nuclei, or lack of migration of nuclei. For example, the chanterelle genus Craterellus often has 6-spored basidia, while some corticioid Sistotrema species can have 2-, 4-, 6-, or 8-spored basidia, and the cultivated button mushroom, Agaricus bisporus can have 1-, 2-, 3- or 4-spored basidia under some circumstances.

Occasionally monokaryons of some taxa can form morphologically fully formed basidiomes and anatomically correct basidia and ballistic basidiospores in the absence of dikaryon formation, diploid nuclei, and meiosis. A rare few number of taxa have extended diploid life cycles, but can be common species. Examples exist in the mushroom genera Armillaria and Xerula, both in the Physalacriaceae.

Occasionally, basidiospores are not formed and parts of the "basidia" act as the dispersal agents, such as with the peculiar mycoparasitic jelly fungus, Tetragoniomyces.

In other cases, the entire "basidium" acts as a "spore," such as in some false puffballs (Scleroderma).

In the human pathogenic genus Filobasidiella, four nuclei following meiosis remain in the basidium but continually divide mitotically, each nucleus migrating into synchronously forming nonballistic basidiospores, which are then pushed upwards by another set forming below them, resulting in 4 parallel chains of dry "basidiospores."

Other variations occur, some as standard life cycles (which themselves have variations within variations) within specific orders.

Rusts

Rusts are fungi of the order Uredinales (or Pucciniales). Many of these species are plant parasites. Many of the rusts have two or more hosts (heteroecious) and up to five spore stages. However, they most commonly reproduce via asexual spore production. Their spores are airborne and can travel great distances. They mostly cause foliar infections. The group received its common name from the fact that some species have a reddish spore stage, which resembles the corrosion process known as rust. Rust occurs on many species of plant, but in most cases any one species of rust can only infect one species of plant.

Rusts at their greatest complexity produce five different types of spores on two different hosts in two unrelated host families. Such rusts are heteroecious (requiring two hosts) and macrocyclic (producing all five spores types). Wheat stem rust is an example.

By convention, the stages and spore states are numbered by Roman numerals. Typically, basidiospores infect host one and the mycelium forms pycnidia, called spermagonia, which are miniature, flask-shaped, hollow, submicroscopic bodies embedded in host tissue (such as a leaf). This stage, numbered "0," produces single-celled, minute spores that ooze out in a sweet liquid and that act as nonmotile spermatia and also protruding receptive hyphae. Insects and probably other vectors, such as rain, carry the spermatia from spermagonia to spermagonia, cross inoculating the mating types. Neither thallus is male or female. Once crossed, the dikaryons are established and a second spore stage is formed, numbered "I" and called aecia, which form dikaryotic aeciospores in dry chains in inverted cup-shaped bodies embedded in host tissue. These aeciospores then infect the second host genus and cannot infect the host on which they are formed (in macrocyclic rusts). On the second host, a repeating spore stage is formed, numbered "II," the uredospores in dry pustules called uredinia. Urediospores are dikaryotic and can infect the same host that produced them. They repeatedly infect this host over the growing season. At the end of the season, a fourth spore type, the teliospore, is formed. It is thicker-walled and serves to overwinter or to survive other harsh conditions. It does not continue the infection process, rather it remains dormant for a period and then germinates to form basidia (stage "IV"), sometimes called a promycelium.

In the Pucciniales (or Uredinales), the basidia are cylindrical and become 3-septate after meiosis, with each of the four cells bearing one basidiospore each. The basidospores disperse and start the infection process on host one again. Autoecious rusts complete their life cycles on one host instead of two, and microcyclic rusts cut out one or more stages.

Smuts

Smuts (mostly Ustilaginomycetes of the class Teliomycetae) cause plant disease, and commonly affect grasses, notably including cereal crops such as maize. They initially attack the plant's reproductive system, forming galls which darken and burst, releasing fungal spores which infect other plants nearby.

The characteristic part of the life cycle of smuts is the thick-walled, often darkly pigmented, ornate, teliospore, which serves to survive harsh conditions such as overwintering and also serves to help disperse the fungus as dry diaspores. The teliospores are initially dikaryotic but become diploid via karyogamy. Meiosis takes place at the time of germination. A promycelim is formed that consists to a short hypha (equated to a basidium).

In some smuts, such as Ustilago maydis, the nuclei migrate into the promycelium that becomes septate, and haploid yeast-like conidia/basidiospores (sometimes called sporidia) bud off laterally from each cell. In various smuts, the yeast phase may proliferate, or they may fuse, or they may infect plant tissue and become hyphal. In other smuts, such as Tilletia caries, the elongated haploid basidiospores form apically, often in compatible pairs that fuse centrally resulting in "H"-shaped diaspores, which are by then dikaryotic. Dikaryotic conidia may then form. Eventually, the host is infected by infectious hyphae. Teliospores form in host tissue. Many variations on these general themes occur.

Smuts with both a yeast phase and an infectious hyphal stage are examples of dimorphic Basidiomycota. In plant parasitic taxa, the saprotrophic phase is normally the yeast, while the infectious stage is hyphal. However, there are examples of animal and human parasites where the species are dimorphic but it is the yeast-like state that is infectious. The genus Filobasidiella forms basidia on hyphae, but the main infectious stage is more commonly known by the anamorphic yeast name Cryptococcus.

The dimorphic Basidiomycota with yeast stages and the pleiomorphic rusts are examples of fungi with anamorphs, which are the asexual stages. Some Basidiomycota are only known as anamorphs. Many are yeasts, collectively called basidiomycetous yeasts to differentiate them from ascomycetous yeasts in the Ascomycota. Aside from yeast anamorphs, and uredinia, aecia, and pycnidia, some Basidiomycota form other distinctive anamorphs as parts of their life cycles. Examples are Collybia tuberosa, with its apple-seed-shaped and colored sclerotium; Dendrocollybia racemosa, with its sclerotium and its Tilachlidiopsis racemosa conidia; Armillaria, with their rhizomorphs; Hohenbuehelia, with their Nematoctonus nematode infectious, state; and the coffee leaf parasite, Mycena citricolor and its Decapitatus flavidus propagules called gemmae.

ASCOMYCOTA

Ascomycota division or phylum of the kingdom Fungi that, together with the Basidiomycota, form the subkingdom Dikarya. Its members are commonly known as the sac fungi or ascomycetes. It is the largest phylum of Fungi, with over 64,000 species. The defining feature of this fungal group is the "ascus", meaning "sac" or "wineskin"), a microscopic sexual structure in which nonmotile spores, called ascospores, are formed. However, some species of the Ascomycota are asexual, meaning that they do not have a sexual cycle and thus do not form asci or ascospores. Familiar examples of sac fungi include morels, truffles, brewer's yeast and baker's yeast, dead man's fingers, and cup fungi. The fungal symbionts in the majority of lichens (loosely termed "ascolichens") such as *Cladonia* belong to the Ascomycota.

Ascomycota is a monophyletic group (it contains all descendants of one common ancestor). Previously placed in the Deuteromycota along with asexual species from other fungal taxa, asexual (or anamorphic) ascomycetes are now identified and classified based on morphological or physiological similarities to ascus-bearing taxa, and by phylogenetic analyses of DNA sequences.

The ascomycetes are of particular use to humans as sources of medicinally important compounds, such as antibiotics, for fermenting bread, alcoholic beverages and cheese. *Penicillium* species on cheeses and those producing antibiotics for treating bacterial infectious diseases are examples of ascomycetes.

Many ascomycetes are pathogens, both of animals, including humans, and of plants. Examples of ascomycetes that can cause infections in humans include *Candida albicans*, *Aspergillus niger* and several tens of species that cause skin infections. The many plant-pathogenic ascomycetes include apple scab, rice blast, the ergot fungi, black knot, and the powdery mildews.

Asexual Reproduction in Ascomycetes and their Characteristics

Ascomycetes

Ascomycetes are 'spore shooters'. They are fungi which produce microscopic spores inside special, elongated cells or sacs, known as 'asci', which give the group its name.

Asexual Reproduction

Asexual reproduction is the dominant form of propagation in the Ascomycota, and is responsible for the rapid spread of these fungi into new areas. Asexual reproduction of ascomycetes is very diverse from both structural and functional points of view. The most important and general is production of conidia, but chlamydospores are also frequently produced. Furthermore, Ascomycota also reproduce asexually through budding.

Conidia Formation

Asexual reproduction may occur through vegetative reproductive spores, the conidia. The asexual, non-motile haploid spores of a fungus, which are named after the Greek word for dust (conia), are hence also known as conidiospores and mitospores. The conidiospores commonly contain one nucleus and are products of mitotic cell divisions and thus are sometimes call mitospores, which are genetically identical to the mycelium from which they originate. They are typically formed at the ends of specialized hyphae, the conidiophores. Depending on the species they may be dispersed by wind or water, or by animals. Conidiophores may simply branch off from the mycelia or they may be formed in fruiting bodies.

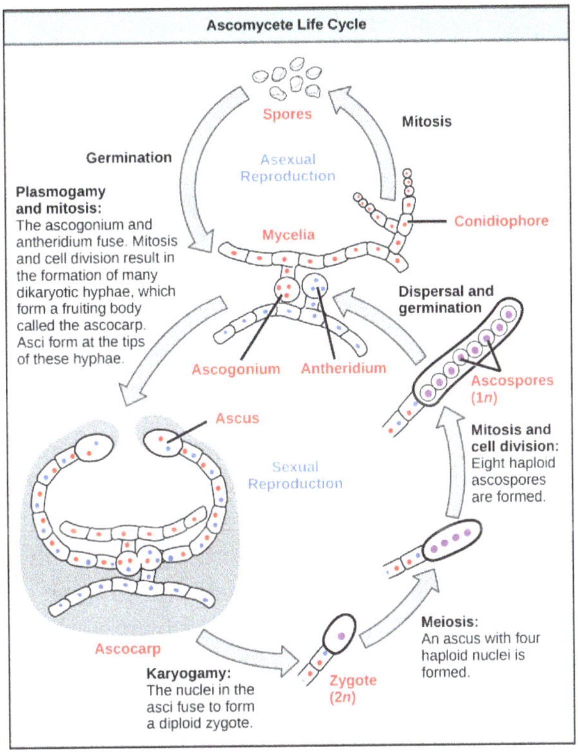

The hypha that creates the sporing (conidiating) tip can be very similar to the normal hyphal tip, or it can be differentiated. The most common differentiation is the formation of a bottle shaped cell called a phialide, from which the spores are produced. Not all of these asexual structures are a single hypha. In some groups, the conidiophores (the structures that bear the conidia) are aggregated to form a thick structure.

E.g., in the order *Moniliales,* all of them are single hyphae with the exception of the aggregations, termed as coremia or synnema. These produce structures rather like cornstokes, with many conidia being produced in a mass from the aggregated conidiophores.

The diverse conidia and conidiophores sometimes develop in asexual sporocarps with different characteristics (e.g. acervulus, pycnidium, sporodochium). Some species of *Ascomycetes* form their structures within plant tissue, either as parasite or saprophytes.

These fungi have evolved more complex asexual sporing structures, probably influenced by the cultural conditions of plant tissue as a substrate. These structures are called the sporodochium. This is a cushion of conidiophores created from a pseudoparenchymatous stroma in plant tissue. The pycnidium is a globose to flask-shaped parenchymatous structure, lined on its inner wall with conidiophores. The acervulus is a flat saucer shaped bed of conidiophores produced under a plant cuticle, which eventually erupt through the cuticle for dispersal.

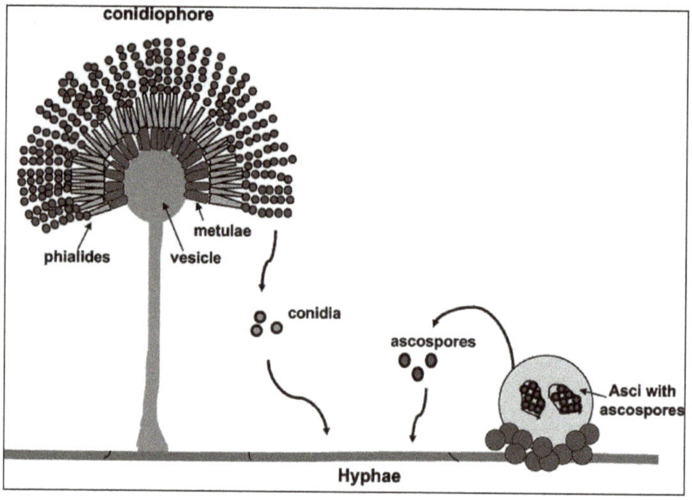

Budding

Asexual reproduction process in ascomycetes also involves the budding which we clearly observe in yeast. This is termed a "blastic process". It involves the blowing out or blebbing of the hyphal tip wall. The blastic process can involve all wall layers, or there can be a new cell wall synthesized which is extruded from within the old wall.

The initial events of budding can be seen as the development of a ring of chitin around the point where the bud is about to appear. This reinforces and stabilizes the cell wall. Enzymatic activity and turgor pressure act to weaken and extrude the cell wall. New cell wall material is incorporated during this phase. Cell contents are forced into the progeny cell, and as the final phase of mitosis ends a cell plate, the point at which a new cell wall will grow inwards from, forms.

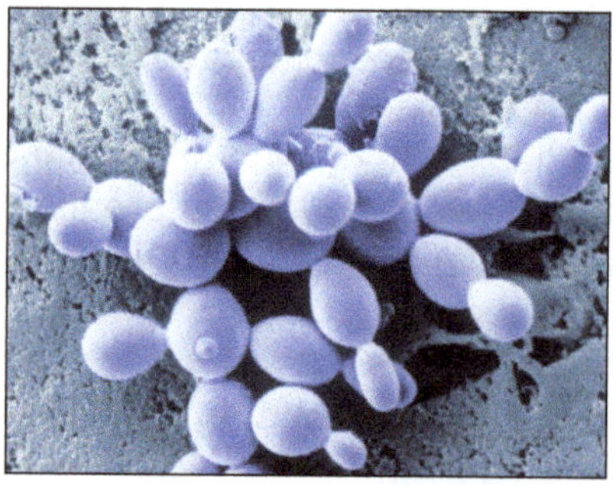

Characteristics of Ascomycetes

- Ascomycota are morphologically diverse. The group includes organisms from unicellular yeasts to complex cup fungi.
- There are 2000 identified genera and 30,000 species of Ascomycota.
- The unifying characteristic among these diverse groups is the presence of a reproductive structure known as the ascus, though in some cases it has a reduced role in the life cycle.
- Many ascomycetes are of commercial importance. Some play a beneficial role, such as the yeasts used in baking, brewing, and wine fermentation, plus truffles and morels, which are held as gourmet delicacies.
- Many of them cause tree diseases, such as Dutch elm disease and apple blights.
- Some of the plant pathogenic ascomycetes are apple scab, rice blast, the ergot fungi, black knot, and the powdery mildews.
- The yeasts are used to produce alcoholic beverages and breads. The mold *Penicillium* is used to produce the antibiotic penicillin.
- Almost half of all members of the phylum Ascomycota form symbiotic associations with algae to form lichens.
- Others, such as morels (a highly prized edible fungi), form important mycorrhizal relationships with plants, thereby providing enhanced water and nutrient uptake and, in some cases, protection from insects.
- Almost all ascomycetes are terrestrial or parasitic. However, a few have adapted to marine or freshwater environments.
- The cell walls of the hyphae are variably composed of chitin and β-glucans, just as in Basidiomycota. However, these fibers are set in a matrix of glycoprotein containing the sugars galactose and mannose.
- The mycelium of ascomycetes is usually made up of septate hyphae. However, there is not necessarily any fixed number of nuclei in each of the divisions.
- The septal walls have septal pores which provide cytoplasmic continuity throughout the individual hyphae. Under appropriate conditions, nuclei may also migrate between septal compartments through the septal pores.
- A unique character of the Ascomycota (but not present in all ascomycetes) is the presence of Woronin bodies on each side of the septa separating the hyphal segments which control the septal pores. If an adjoining hypha is ruptured, the Woronin bodies block the pores to prevent loss of cytoplasm into the ruptured

compartment. The Woronin bodies are spherical, hexagonal, or rectangular membrane bound structures with a crystalline protein matrix.

Modern Classification

There are three subphyla that are described and accepted:

- The Pezizomycotina are the largest subphylum and contains all ascomycetes that produce ascocarps (fruiting bodies), except for one genus, Neolecta, in the Taphrinomycotina. It is roughly equivalent to the previous taxon, Euascomycetes. The Pezizomycotina includes most macroscopic "ascos" such as truffles, ergot, ascolichens, cup fungi (discomycetes), pyrenomycetes, lorchels, and caterpillar fungus. It also contains microscopic fungi such as powdery mildews, dermatophytic fungi, and Laboulbeniales.
- The Saccharomycotina comprise most of the "true" yeasts, such as baker's yeast and Candida, which are single-celled (unicellular) fungi, which reproduce vegetatively by budding. Most of these species were previously classified in a taxon called Hemiascomycetes.
- The Taphrinomycotina include a disparate and basal group within the Ascomycota that was recognized following molecular (DNA) analyses. The taxon was originally named Archiascomycetes (or Archaeascomycetes). It includes hyphal fungi (Neolecta, Taphrina, Archaeorhizomyces), fission yeasts (Schizosaccharomyces), and the mammalian lung parasite Pneumocystis.

Outdated Taxon Names

Several outdated taxon names—based on morphological features—are still occasionally used for species of the Ascomycota. These include the following sexual (teleomorphic) groups, defined by the structures of their sexual fruiting bodies: The Discomycetes, which included all species forming apothecia; the Pyrenomycetes, which included all sac fungi that formed perithecia or pseudothecia, or any structure resembling these morphological structures; and the Plectomycetes, which included those species that form cleistothecia. Hemiascomycetes included the yeasts and yeast-like fungi that have now been placed into the Saccharomycotina or Taphrinomycotina, while the Euascomycetes included the remaining species of the Ascomycota, which are now in the Pezizomycotina, and the Neolecta, which are in the Taphrinomycotina.

Some ascomycetes do not reproduce sexually or are not known to produce asci and are therefore anamorphic species. Those anamorphs that produce conidia (mitospores) were previously described as mitosporic Ascomycota. Some taxonomists placed this group into a separate artificial phylum, the Deuteromycota (or "Fungi Imperfecti"). Where recent molecular analyses have identified close relationships with ascus-bearing taxa, anamorphic species have been grouped into the Ascomycota, despite the absence of the defining ascus. Sexual and asexual isolates of the same species commonly carry different binomial species names, as, for example, *Aspergillus nidulans* and *Emericella nidulans*, for asexual and sexual isolates, respectively, of the same species.

Species of the Deuteromycota were classified as Coelomycetes if they produced their conidia in minute flask- or saucer-shaped conidiomata, known technically as *pycnidia* and *acervuli*. The Hyphomycetes were those species where the conidiophores (*i.e.*, the hyphal structures that carry conidia-forming cells at the end) are free or loosely organized. They are mostly isolated but sometimes also appear as bundles of cells aligned in parallel (described as *synnematal*) or as cushion-shaped masses (described as *sporodochial*).

Morphology

A member of the genus *Cordyceps* which is parasitic on arthropods. Note the elongated stromata. Species unknown, perhaps *Cordyceps ignota*.

Most species grow as filamentous, microscopic structures called hyphae or as budding single cells (yeasts). Many interconnected hyphae form a thallus usually referred to as the mycelium, which—when visible to the naked eye (macroscopic)—is commonly called mold. During sexual reproduction, many Ascomycota typically produce large numbers of asci. The ascus is often contained in a multicellular, occasionally readily visible fruiting structure, the ascocarp (also called an *ascoma*). Ascocarps come in a very large variety of shapes: cup-shaped, club-shaped, potato-like, spongy, seed-like, oozing and pimple-like, coral-like, nit-like, golf-ball-shaped, perforated tennis ball-like, cushion-shaped, plated and feathered in miniature (Laboulbeniales), microscopic classic Greek shield-shaped, stalked or sessile. They can appear solitary or clustered. Their texture can likewise be very variable, including fleshy, like charcoal (carbonaceous), leathery, rubbery, gelatinous, slimy, powdery, or cob-web-like. Ascocarps come in multiple colors such as red, orange, yellow, brown, black, or, more rarely, green or blue. Some ascomyceous fungi, such as *Saccharomyces cerevisiae*, grow as single-celled yeasts, which—during sexual reproduction—develop into an ascus, and do not form fruiting bodies.

The "candlesnuff fungus" in its asexual state, *Xylaria hypoxylon*.

In lichenized species, the thallus of the fungus defines the shape of the symbiotic colony. Some dimorphic species, such as *Candida albicans*, can switch between growth as single cells and as filamentous, multicellular hyphae. Other species are pleomorphic, exhibiting asexual (anamorphic) as well as a sexual (teleomorphic) growth forms.

The ascocarp of a morel contains numerous apothecia.

Except for lichens, the non-reproductive (vegetative) mycelium of most ascomycetes is usually inconspicuous because it is commonly embedded in the substrate, such as soil, or grows on or inside a living host, and only the ascoma may be seen when fruiting. Pigmentation, such as melanin in hyphal walls, along with prolific growth on surfaces can result in visible mold colonies; examples include *Cladosporium* species, which form black spots on bathroom caulking and other moist areas. Many ascomycetes cause food spoilage, and, therefore, the pellicles or moldy layers that develop on jams, juices, and other foods are the mycelia of these species or occasionally Mucoromycotina and almost never Basidiomycota. Sooty molds that develop on plants, especially in the tropics are the thalli of many species.

Large masses of yeast cells, asci or ascus-like cells, or conidia can also form macroscopic structures. For example. *Pneumocystis* species can colonize lung cavities (visible in x-rays), causing a form of pneumonia. Asci of *Ascosphaera* fill honey bee larvae and pupae causing mummification with a chalk-like appearance, hence the name "chalkbrood". Yeasts for small colonies in vitro and in vivo, and excessive growth of *Candida* species in the mouth or vagina causes "thrush", a form of candidiasis.

The cell walls of the ascomycetes almost always contain chitin and β-glucans, and divisions within the hyphae, called "septa", are the internal boundaries of individual cells (or compartments). The cell wall and septa give stability and rigidity to the hyphae and may prevent loss of cytoplasm in case of local damage to cell wall and cell membrane. The septa commonly have a small opening in the center, which functions as a cytoplasmic connection between adjacent cells, also sometimes allowing cell-to-cell movement of nuclei within a hypha. Vegetative hyphae of most ascomycetes contain only one nucleus per cell (*uninucleate* hyphae), but multinucleate cells—especially in the apical regions of growing hyphae—can also be present.

Metabolism

In common with other fungal phyla, the Ascomycota are heterotrophic organisms that require organic compounds as energy sources. These are obtained by feeding on a variety of organic substrates including dead matter, foodstuffs, or as symbionts in or on other living organisms. To obtain these nutrients from their surroundings, ascomycetous fungi secrete powerful digestive enzymes that break down organic substances into smaller molecules, which are then taken up into the cell. Many species live on dead plant material such as leaves, twigs, or logs. Several species colonize plants, animals, or other fungi as parasites or mutualistic symbionts and derive all their metabolic energy in form of nutrients from the tissues of their hosts.

Owing to their long evolutionary history, the Ascomycota have evolved the capacity to break down almost every organic substance. Unlike most organisms, they are able to use their own enzymes to digest plant biopolymers such as cellulose or lignin. Collagen, an abundant structural protein in animals, and keratin—a protein that forms hair and nails—, can also serve as food sources. Unusual examples include *Aureobasidium pullulans*, which feeds on wall paint, and the kerosene fungus *Amorphotheca resinae*, which feeds on aircraft fuel (causing occasional problems for the airline industry), and may sometimes block fuel pipes. Other species can resist high osmotic stress and grow, for example, on salted fish, and a few ascomycetes are aquatic.

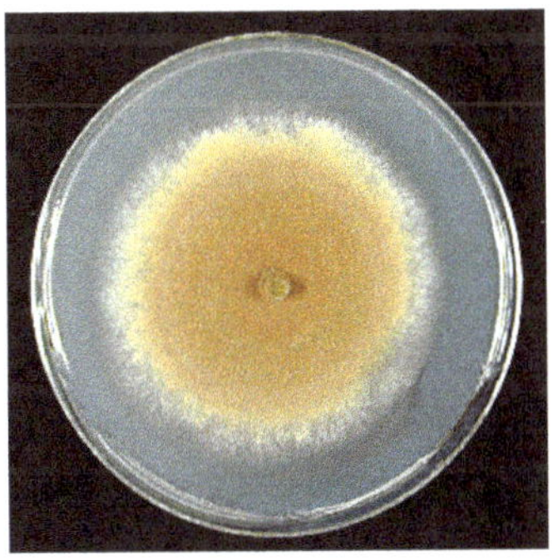

Hypomyces completus on culture medium.

The Ascomycota is characterized by a high degree of specialization; for instance, certain species of Laboulbeniales attack only one particular leg of one particular insect species.

Many Ascomycota engage in symbiotic relationships such as in lichens—symbiotic associations with green algae or cyanobacteria—in which the fungal symbiont directly obtains products of photosynthesis. In common with many basidiomycetes and Glomeromycota, some ascomycetes form symbioses with plants by colonizing the roots to form mycorrhizal associations. The Ascomycota also represents several carnivorous fungi, which have developed hyphal traps to capture small protists such as amoebae, as well as roundworms (*Nematoda*), rotifers, tardigrades, and small arthropods such as springtails (*Collembola*).

Distribution and Living Environment

The Ascomycota are represented in all land ecosystems worldwide, occurring on all continents including Antarctica. Spores and hyphal fragments are dispersed through the atmosphere and freshwater environments, as well as ocean beaches and tidal zones. The distribution of species is variable; while some are found on all continents, others, as for example the white truffle *Tuber magnatum*, only occur in isolated locations in Italy and Eastern Europe. The distribution of plant-parasitic species is often restricted by host distributions; for example, *Cyttaria* is only found on *Nothofagus* (Southern Beech) in the Southern Hemisphere.

Reproduction

Asexual Reproduction

Asexual reproduction is the dominant form of propagation in the Ascomycota, and is responsible for the rapid spread of these fungi into new areas. It occurs through vegetative reproductive spores, the conidia. The conidiospores commonly contain one nucleus and are products of mitotic cell divisions and thus are sometimes called mitospores, which are genetically identical to the mycelium from which they originate. They are typically formed at the ends of specialized hyphae, the *conidiophores*. Depending on the species they may be dispersed by wind or water, or by animals.

Asexual Spores

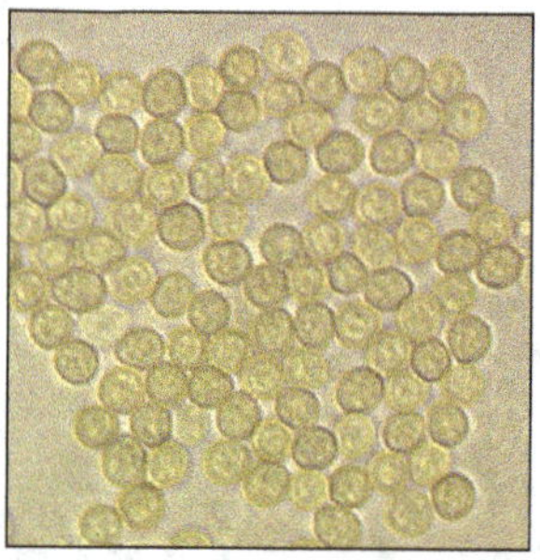

Conidiospores of Trichoderma aggressivum, Diameter approx. 3µm.

Different types of asexual spores can be identified by colour, shape, and how they are released as individual spores. Spore types can be used as taxonomic characters in the classification within the Ascomycota. The most frequent types are the single-celled spores, which are designated *amerospores*. If the spore is divided into two by a cross-wall (septum), it is called a *didymospore*.

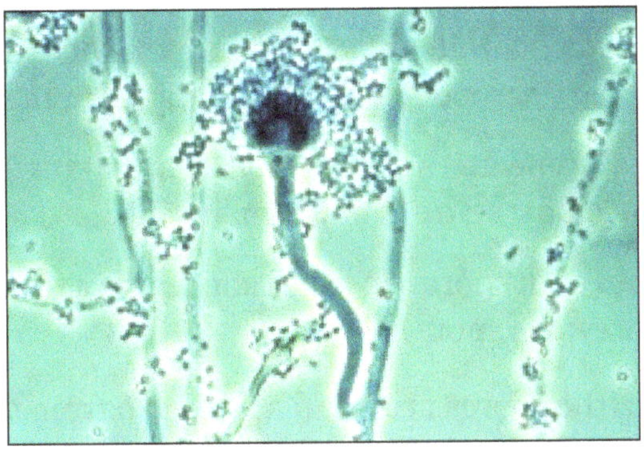

Conidiophores of molds of the genus Aspergillus, conidiogenesis is blastic-phialidic.

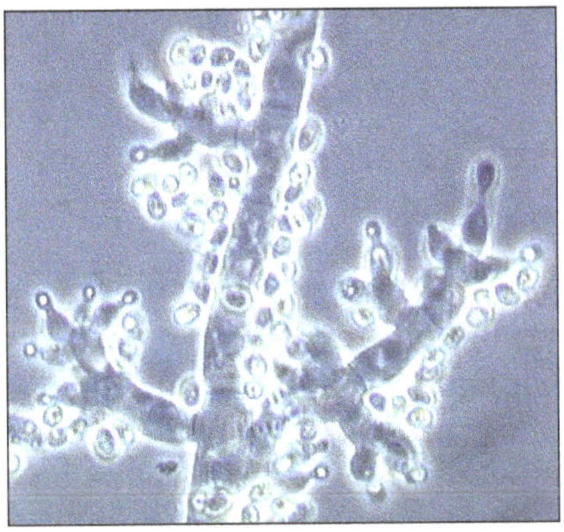

Conidiophores of Trichoderma fertile with vase-shaped phialides and newly formed conidia on their ends (bright points).

When there are two or more cross-walls, the classification depends on spore shape. If the septae are *transversal*, like the rungs of a ladder, it is a *phragmospore*, and if they possess a net-like structure it is a *dictyospore*. In *staurospores* ray-like arms radiate from a central body; in others (*helicospores*) the entire spore is wound up in a spiral like a spring. Very long worm-like spores with a length-to-diameter ratio of more than 15:1, are called *scolecospores*.

Conidiogenesis and Dehiscence

Important characteristics of the anamorphs of the Ascomycota are *conidiogenesis*, which includes spore formation and dehiscence (separation from the parent structure). Conidiogenesis corresponds to Embryology in animals and plants and can be divided into two fundamental forms of development: *blastic* conidiogenesis, where the spore is already evident before it separates from the conidiogenic hypha, and *thallic* conidiogenesis, during which a cross-wall forms and the newly created cell develops into a spore. The spores may or may not be generated in a large-scale specialized structure that helps to spread them.

These two basic types can be further classified as follows:

- Blastic-acropetal (repeated budding at the tip of the conidiogenic hypha, so that a chain of spores is formed with the youngest spores at the tip).
- Blastic-synchronous (simultaneous spore formation from a central cell, sometimes with secondary acropetal chains forming from the initial spores).
- Blastic-sympodial (repeated sideways spore formation from behind the leading spore, so that the oldest spore is at the main tip).
- Blastic-annellidic (each spore separates and leaves a ring-shaped scar inside the scar left by the previous spore).
- Blastic-phialidic (the spores arise and are ejected from the open ends of special conidiogenic cells called phialides, which remain constant in length).
- Basauxic (where a chain of conidia, in successively younger stages of development, is emitted from the mother cell).
- Blastic-retrogressive (spores separate by formation of crosswalls near the tip of the conidiogenic hypha, which thus becomes progressively shorter).
- Thallic-arthric (double cell walls split the conidiogenic hypha into cells that develop into short, cylindrical spores called arthroconidia; sometimes every second cell dies off, leaving the arthroconidia free).
- Thallic-solitary (a large bulging cell separates from the conidiogenic hypha, forms internal walls, and develops to a phragmospore).

Sometimes the conidia are produced in structures visible to the naked eye, which help to distribute the spores. These structures are called "conidiomata" (singular: conidioma), and may take the form of *pycnidia* (which are flask-shaped and arise in the fungal tissue) or *acervuli* (which are cushion-shaped and arise in host tissue).

Dehiscence happens in two ways. In *schizolytic* dehiscence, a double-dividing wall with a central lamella (layer) forms between the cells; the central layer then breaks down thereby releasing the spores. In *rhexolytic* dehiscence, the cell wall that joins the spores on the outside degenerates and releases the conidia.

Heterokaryosis and Parasexuality

Several Ascomycota species are not known to have a sexual cycle. Such asexual species may be able to undergo genetic recombination between individuals by processes involving *heterokaryosis* and *parasexual* events.

Parasexuality refers to the process of heterokaryosis, caused by merging of two hyphae belonging to different individuals, by a process called *anastomosis*, followed by a series of events resulting in genetically different cell nuclei in the mycelium. The merging of nuclei is not followed by meiotic events, such as gamete formation and results in an increased number of chromosomes per nuclei. *Mitotic crossover* may enable recombination, i.e., an exchange of genetic material between homologous chromosomes. The chromosome number may then be restored to its haploid state by nuclear division, with each daughter nuclei being genetically different from the original parent nuclei. Alternatively, nuclei may lose some chromosomes, resulting in aneuploid cells. *Candida albicans* (class Saccharomycetes) is an example of a fungus that has a parasexual cycle

Sexual Reproduction

Sexual reproduction in the Ascomycota leads to the formation of the *ascus*, the structure that defines this fungal group and distinguishes it from other fungal phyla. The ascus is a tube-shaped vessel, a *meiosporangium*, which contains the sexual spores produced by meiosis and which are called *ascospores*.

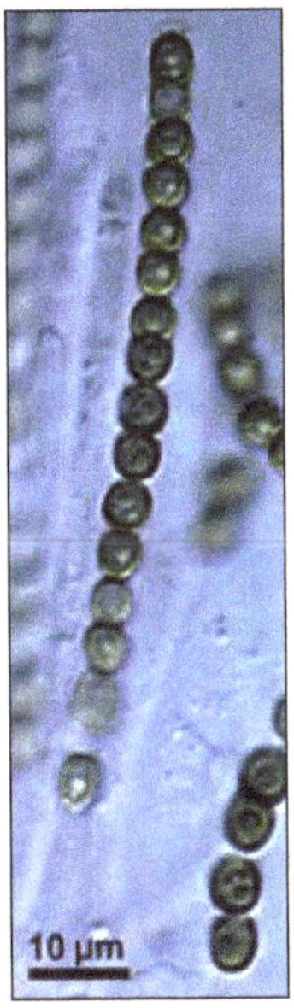

Ascus of *Hypocrea virens* with eight two-celled Ascospores.

Apart from a few exceptions, such as *Candida albicans*, most ascomycetes are haploid, i.e., they contain one set of chromosomes per nucleus. During sexual reproduction there is a diploid phase, which commonly is very short, and meiosis restores the haploid state. The sexual cycle of one well-studied representative species of Ascomycota is described in greater detail in *Neurospora crassa*. Also, the adaptive basis for the maintenance of sexual reproduction in the Ascomycota fungi was reviewed by Wallen and Perlin. They concluded that the most plausible reason for the maintenance of this capability is the benefit of repairing DNA damage by using recombination that occurs during meiosis. DNA damage can be caused by a variety of stresses such as nutrient limitation.

Formation of Sexual Spores

The sexual part of the life cycle commences when two hyphal structures mate. In the case of *homothallic* species, mating is enabled between hyphae of the same fungal clone, whereas in *heterothallic* species, the two hyphae must originate from fungal clones that differ genetically, i.e., those that are of a different mating type. Mating types are typical of the fungi and correspond roughly to the sexes in plants and animals; however one species may have more than two mating types, resulting in sometimes complex vegetative incompatibility systems.

Gametangia are sexual structures formed from hyphae, and are the generative cells. A very fine hypha, called trichogyne emerges from one gametangium, the *ascogonium*, and merges with a gametangium (the *antheridium*) of the other fungal isolate. The nuclei in the antheridium then migrate into the ascogonium, and plasmogamy—the mixing of the cytoplasm—occurs. Unlike in animals and plants, plasmogamy is not immediately followed by the merging of the nuclei (called *karyogamy*). Instead, the nuclei from the two hyphae form pairs, initiating the *dikaryophase* of the sexual cycle, during which time the pairs of nuclei synchronously divide. Fusion of the paired nuclei leads to mixing of the genetic material and recombination and is followed by meiosis. A similar sexual cycle is present in the red algae (Rhodophyta). A discarded hypothesis held that a second karyogamy event occurred in the ascogonium prior to ascogeny, resulting in a tetraploid nucleus which divided into four diploid nuclei by meiosis and then into eight haploid nuclei by a supposed process called brachymeiosis, but this hypothesis was disproven in the 1950s.

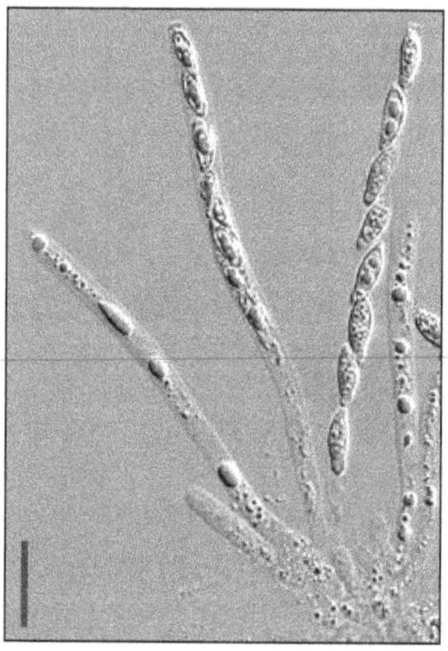

Unitunicate-inoperculate Asci of *Hypomyces chrysospermus*.

From the fertilized ascogonium, *dinucleate* hyphae emerge in which each cell contains two nuclei. These hyphae are called *ascogenous* or fertile hyphae. They are supported by the vegetative mycelium containing uni– (or mono–) nucleate hyphae, which are sterile. The mycelium containing both sterile and fertile hyphae may grow into fruiting body, the *ascocarp*, which may contain millions of fertile hyphae.

An ascocarp is the fruiting body of the sexual phase in Ascomycota. There are five morphologically different types of ascocarp, namely:

- Naked asci: These are simple ascomycetes and asci are produced on the plants surface.
- Perithecia: Asci are in flask-shaped ascoma (perithecium) with a pore (ostiole) at the top.
- Cleistothecia: The ascocarp (a cleistothecium) is spherical and closed.
- Apothecia: The asci are in a bowl shaped ascoma (apothecium). These are sometimes called the "cup fungi".
- Pseudothecia: Asci with two layers, produced in pseudothecia that look like perithecia. The ascospores are arranged irregularly.

The sexual structures are formed in the fruiting layer of the ascocarp, the hymenium. At one end of ascogenous hyphae, characteristic U-shaped hooks develop, which curve back opposite to the growth direction of the hyphae. The two nuclei contained in the apical part of each hypha divide in such a way that the threads of their mitotic spindles run parallel, creating two pairs of genetically different nuclei. One daughter nucleus migrates close to the hook, while the other daughter nucleus locates to the basal part of the hypha. The formation of two parallel cross-walls then divides the hypha into three sections: one at the hook with one nucleus, one at the basal of the original hypha that contains one nucleus, and one that separates the U-shaped part, which contains the other two nuclei.

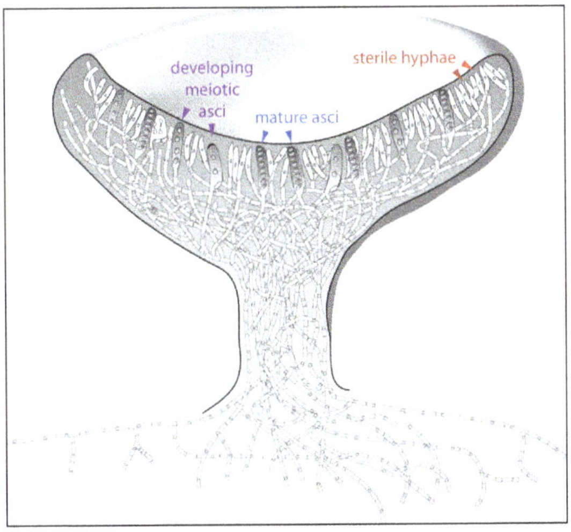

Diagram of an apothecium (the typical cup-like reproductive structure of Ascomycetes) showing sterile tissues as well as developing and mature asci.

Fusion of the nuclei (karyogamy) takes place in the U-shaped cells in the hymenium, and results in the formation of a diploid zygote. The zygote grows into the ascus, an elongated tube-shaped or cylinder-shaped capsule. Meiosis then gives rise to four haploid nuclei, usually followed by a further mitotic division that results in eight nuclei in each ascus. The nuclei along with some cytoplasma become enclosed within membranes and a cell wall to give rise to ascospores that are aligned inside the ascus like peas in a pod.

Upon opening of the ascus, ascospores may be dispersed by the wind, while in some cases the spores are forcibly ejected form the ascus; certain species have evolved spore cannons, which can eject ascospores up to 30 cm. away. When the spores reach a suitable substrate, they germinate, form new hyphae, which restarts the fungal life cycle.

The form of the ascus is important for classification and is divided into four basic types: unitunicate-operculate, unitunicate-inoperculate, bitunicate, or prototunicate.

Ecology

The Ascomycota fulfil a central role in most land-based ecosystems. They are important decomposers, breaking down organic materials, such as dead leaves and animals, and helping the detritivores (animals that feed on decomposing material) to obtain their nutrients. Ascomycetes along with other fungi can break down large molecules such as cellulose or lignin, and thus have important roles in nutrient cycling such as the carbon cycle.

The fruiting bodies of the Ascomycota provide food for many animals ranging from insects and slugs and snails (*Gastropoda*) to rodents and larger mammals such as deer and wild boars.

Many ascomycetes also form symbiotic relationships with other organisms, including plants and animals.

Lichens

Probably since early in their evolutionary history, the Ascomycota have formed symbiotic associations with green algae (*Chlorophyta*), and other types of algae and cyanobacteria. These mutualistic associations are commonly known as lichens, and can grow and persist in terrestrial regions of the earth that are inhospitable to other organisms and characterized by extremes in temperature and humidity, including the Arctic, the Antarctic, deserts, and mountaintops. While the photoautotrophic algal partner generates metabolic energy through photosynthesis, the fungus offers a stable, supportive matrix and protects cells from radiation and dehydration. Around 42% of the Ascomycota (about 18,000 species) form lichens, and almost all the fungal partners of lichens belong to the Ascomycota.

Mycorrhizal Fungi and Endophytes

Members of the Ascomycota form two important types of relationship with plants: As mycorrhizal fungi and as endophytes. Mycorrhiza are symbiotic associations of fungi with the root systems of the plants, which can be of vital importance for growth and persistence for the plant. The fine mycelial network of the fungus enables the increased uptake of mineral salts that occur at low levels in the soil. In return, the plant provides the fungus with metabolic energy in the form of photosynthetic products.

Endophytic fungi live inside plants, and those that form mutualistic or commensal associations with their host, do not damage their hosts. The exact nature of the relationship between endophytic fungus and host depends on the species involved, and in some cases fungal colonization of plants can bestow a higher resistance against insects, roundworms (nematodes), and bacteria; in the case of grass endophytes the fungal symbiont produces poisonous alkaloids, which can affect the health of plant-eating (herbivorous) mammals and deter or kill insect herbivores.

Symbiotic Relationships with Animals

Several ascomycetes of the genus Xylaria colonize the nests of leafcutter ants and other fungus-growing ants of the tribe Attini, and the fungal gardens of termites (Isoptera). Since they do not generate fruiting bodies until the insects have left the nests, it is suspected that, as confirmed in several cases of Basidiomycota species, they may be cultivated.

Bark beetles (family Scolytidae) are important symbiotic partners of ascomycetes. The female beetles transport fungal spores to new hosts in characteristic tucks in their skin, the *mycetangia*. The beetle tunnels into the wood and into large chambers in which they lay their eggs. Spores released from the mycetangia germinate into hyphae, which can break down the wood. The beetle larvae then feed on the fungal mycelium, and, on reaching maturity, carry new spores with them to renew the cycle of infection. A well-known example of this is Dutch elm disease, caused by Ophiostoma ulmi, which is carried by the European elm bark beetle, Scolytus multistriatus.

Plant Disease Interactions

One of their most harmful roles is as the agent of many plant diseases. For instance:

- Dutch Elm Disease, caused by the closely related species Ophiostoma ulmi and Ophiostoma novo-ulmi, has led to the death of many elms in Europe and North America.
- The originally Asian Cryphonectria parasitica is responsible for attacking Sweet Chestnuts (Castanea sativa), and virtually eliminated the once-widespread American Chestnut (Castanea dentata).
- A disease of maize (Zea mays), which is especially prevalent in North America, is brought about by Cochliobolus heterostrophus.
- Taphrina deformans causes leaf curl of peach.

- Uncinula necator is responsible for the disease powdery mildew, which attacks grapevines.
- Species of Monilinia cause brown rot of stone fruit such as peaches (Prunus persica) and sour cherries (Prunus ceranus).
- Members of the Ascomycota such as Stachybotrys chartarum are responsible for fading of woolen textiles, which is a common problem especially in the tropics.
- Blue-green, red and brown molds attack and spoil foodstuffs - for instance Penicillium italicum rots oranges.
- Cereals infected with Fusarium graminearum contain mycotoxins like deoxynivalenol (DON), which causes Fusarium ear blight and skin and mucous membrane lesions when eaten by pigs.

Human Disease Interactions

- Aspergillus fumigatus, the most common cause of fungal infection in the lungs of immune compromised patients often resulting in death. Also the most frequent cause of Allergic bronchopulmonary aspergillosis, which often occurs in patients with Cystic fibrosis as well as Asthma.
- Candida albicans, a yeast that attacks the mucous membranes, can cause an infection of the mouth or vagina called thrush or candidiasis, and is also blamed for "yeast allergies".
- Fungi like Epidermophyton cause skin infections but are not very dangerous for people with healthy immune systems. However, if the immune system is damaged they can be life-threatening; for instance, Pneumocystis jirovecii is responsible for severe lung infections that occur in AIDS patients.
- Ergot (Claviceps purpurea) is a direct menace to humans when it attacks wheat or rye and produces highly poisonous alkaloids, causing ergotism if consumed. Symptoms include hallucinations, stomach cramps, and a burning sensation in the limbs ("Saint Anthony's Fire").
- Aspergillus flavus, which grows on peanuts and other hosts, generates aflatoxin, which damages the liver and is highly carcinogenic.

Beneficial effects for Humans

On the other hand, ascus fungi have brought some important benefits to humanity.

- The most famous case may be that of the mould Penicillium chrysogenum (formerly Penicillium notatum), which, probably to attack competing bacteria, produces an antibiotic that, under the name of penicillin, triggered a revolution in the treatment of bacterial infectious diseases in the 20th century.

- The medical importance of Tolypocladium niveum as an immunosuppressor can hardly be exaggerated. It excretes Ciclosporin, which, as well as being given during Organ transplantation to prevent rejection, is also prescribed for auto-immune diseases such as multiple sclerosis, although there is some doubt over the long-term side-effects of the treatment.

Stilton cheese veined with *Penicillium roqueforti*.

- Some ascomycete fungi can be altered relatively easily through genetic engineering procedures. They can then produce useful proteins such as insulin, human growth hormone, or TPa, which is employed to dissolve blood clots.
- Several species are common model organisms in biology, including Saccharomyces cerevisiae, Schizosaccharomyces pombe, and Neurospora crassa. The genomes of a number of ascomycete fungi have been fully sequenced.
- Baker's Yeast (Saccharomyces cerevisiae) is used to make bread, beer and wine, during which process sugars such as glucose or sucrose are fermented to make ethanol and carbon dioxide. Bakers use the yeast for carbon dioxide production, causing the bread to rise, with the ethanol boiling off during cooking. Most vintners use it for ethanol production, with the carbon dioxide being released into the atmosphere during fermentation. Brewers and traditional producers of sparkling wine use both, with a primary fermentation for the alcohol and a secondary one to produce the carbon dioxide bubbles that provide the drinks with "sparkling" texture in the case of wine and the desirable foam in the case of beer.
- Enzymes of Penicillium camemberti play a role in the manufacture of the cheeses Camembert and Brie, while those of Penicillium roqueforti do the same for Gorgonzola, Roquefort and Stilton.
- In Asia, Aspergillus oryzae is added to a pulp of soaked soya beans to make soy sauce, and is used to break down starch in rice and other grains into simple sugars for fermentation into East Asian alcoholic beverages such as huangjiu and sake.

- Finally, some members of the Ascomycota are choice edibles; morels (Morchella spp.), truffles (Tuber spp.), and lobster mushroom (Hypomyces lactifluorum) are some of the most sought-after fungal delicacies.

NEOCALLIMASTIGOMYCOTA

Neocallimastigomycota is a phylum containing anaerobic fungi, which are symbionts found in the digestive tracts of larger herbivores. Anaerobic fungi were originally placed within phylum Chytridiomycota, within Order Neocallimastigales but later raised to phylum level, a decision upheld by later phylogenetic reconstructions. It encompasses only one family.

Discovery

The fungi in Neocallimastigomycota were first recognised as fungi by Orpin in 1975, based on motile cells present in the rumen of sheep. Their zoospores had been observed much earlier but were believed to be flagellate protists, but Orpin demonstrated that they possessed a chitin cell wall. It has since been shown that they are fungi related to the core chytrids. Prior to this, the microbial population of the rumen was believed to consist only of bacteria and protozoa. Since their discovery they have been isolated from the digestive tracts of over 50 herbivores, including ruminant and non-ruminant (hindgut-fermenting) mammals and herbivorous reptiles.

Neocallimastigomycota have also been found in humans.

Reproduction and Growth

These fungi reproduce in the rumen of ruminants through the formation of zoospores which are released from sporangia. These zoospores bear a kinetosome but lack the non-flagellated centriole known in most chytrids, and have been known to utilize horizontal gene transfer in their development of xylanase (from bacteria) and other glucanases.

The nuclear envelopes of their cells are notable for remaining intact throughout mitosis. Sexual reproduction has not been observed in anaerobic fungi. However, they are known to be able to survive for many months in aerobic environments, a factor which is important in they colonisation of new hosts. In *Anaeromyces*, the presence of putative resting spores has been observed but the way in which these are formed and germinate remains unknown.

Metabolism

Neocallimastigomycota lack mitochondria but instead contain hydrogenosomes in which the oxidation of NADH to NAD+, leading to formation of H_2.

Polysaccharide-degrading Activity

Neocallimastigomycota play an essential role in fibre-digestion in their host species. They are present in large numbers in the digestive tracts of animals which are fed on high fibre diets. The polysaccharide degrading enzymes produced by anaerobic fungi can hydrolyse the most recalcitrant plant polymers and can degrade unlignified plant cell walls entirely. The polysaccharide degrading enzymes are organised into a multi-protein complex, similar to the bacterial cellulosome.

BLASTOCLADIOMYCOTA

Blastocladiomycota is one of the currently recognized phyla within the kingdom Fungi. Blastocladiomycota was originally the order Blastocladiales within the phylum Chytridiomycota until molecular and zoospore ultrastructural characters were used to demonstrate it was not monophyletic with Chytridiomycota. The order was first erected by Petersen for a single genus, *Blastocladia*, which was originally considered a member of the oomycetes. Accordingly, members of Blastocladiomycota are often referred to colloquially as "chytrids." However, some feel "chytrid" should refer only to members of Chytridiomycota. Thus, members of Blastocladiomyota are commonly called "blastoclads" by mycologists. Alternatively, members of Blastocladiomycota, Chytridiomycota, and Neocallimastigomycota lumped together as the zoosporic true fungi. Blastocladiomycota contains 5 families and approximately 12 general. This early diverging branch of kingdom Fungi is the first to exhibit alternation of generations. As well, two (once) popular model organisms—*Allomyces macrogynus* and *Blastocladiella emersonii*—belong to this phylum.

Morphology

Morphology in Blastocladiomycota varies greatly. For example, members of Coelomycetaceae are simple, unwalled, and plasmodial in nature. Some species in *Blastocladia* are monocentric, like the chytrids, while others are polycentric. The most remarkable are those members, such as *Allomyces* that demonstrate determinant, differentiated growth.

Reproduction/Life Cycle

Sexual Reproduction

As stated above, some members of Blastocladiomycota exhibit alternation of generations. Members of this phylum also exhibit a form of sexual reproduction known as anisogamy. Anisogamy is the fusion of two sexual gametes that differ in morphology, usually size. In *Allomyces*, the thallus (body) is attached by rhizoids, and has an erect trunk on which reproductive organs are formed at the end of branches. During the haploid phase, the thallus forms male and female gametangia that release flagellated gametes. Gametes attract one another using pheromones and eventually fuse to form a Zygote. The germinated zygote produces a diploid thallus with two types of sporangia: Thin-walled zoosporangia and thick walled resting spores (or sporangia). The thin walled sporangia release diploid zoospores. The resting spore serves as a means of enduring unfavorable conditions. When conditions are favorable again, meiosis occurs and haploid zoospores are released. These germinate and grow into haploid thalli that will produce "male" and "female" gametangia and gametes.

Asexual Reproduction

Similar to Chytridiomycota, members of Blastocladiomycota produce asexual zoospores to colonize new substrates. In some species, a curious phenomenon has been observed in the asexual zoospores. From time to time, asexual zoospores will pair up and exchange cytoplasm but not nuclei.

Ecological Roles

Plant leaf with *Physoderma menyanthis* (former *Cladochytrium menyanthis*) signs.

Similar to Chytridiomycota, members of Blastocladiomycota are capable of growing on refractory materials, such as pollen, keratin, cellulose, and chitin. The best known species, however, are the parasites. Members of *Catenaria* are parasites of nematodes, midges, crustaceans, and even another blastoclad, *Coelomyces*. Members of the genus *Physoderma* and *Urophlyctis* are obligate plant parasites. Of economic importance is *Physoderma maydis*, a parasite of maize and the causal agent of brown spot disease. Also of importance are the species of *Urophlyctis* that parasitize alfalfa. However, ecologically, *Physoderma* are important parasites of many aquatic and marsh angiosperms. Also of human interest, for health reasons, are members of *Coelomomyces*, an unusual parasite of mosquitoes that requires an alternate crustacean host (the same one parasitized by members of *Catenaria*) to complete its life cycle. Others that are ecologically interesting include a parasite of water bears and the zooplankter *Daphnia*.

GLOMEROMYCOTA

Glomeromycota (often referred to as glomeromycetes, as they include only one class, Glomeromycetes) are one of eight currently recognized divisions within the kingdom

Fungi, with approximately 230 described species. Members of the Glomeromycota form arbuscular mycorrhizas (AMs) with the thalli of bryophytes and the roots of vascular land plants. Not all species have been shown to form AMs, and one, *Geosiphon pyriformis*, is known not to do so. Instead, it forms an endocytobiotic association with *Nostoc* cyanobacteria. The majority of evidence shows that the Glomeromycota are dependent on land plants (*Nostoc* in the case of *Geosiphon*) for carbon and energy, but there is recent circumstantial evidence that some species may be able to lead an independent existence. The arbuscular mycorrhizal species are terrestrial and widely distributed in soils worldwide where they form symbioses with the roots of the majority of plant species (>80%). They can also be found in wetlands, including salt-marshes, and associated with epiphytic plants.

Reproduction

The Glomeromycota have generally coenocytic (occasionally sparsely septate) mycelia and reproduce asexually through blastic development of the hyphal tip to produce spores (Glomerospores) with diameters of 80–500 μm. In some, complex spores form within a terminal saccule. Recently it was shown that *Glomus* species contain 51 genes encoding all the tools necessary for meiosis. Based on these and related findings, it was suggested that *Glomus* species may have a cryptic sexual cycle.

Colonization

New colonization of AM fungi largely depends on the amount of inoculum present in the soil. Although pre-existing hyphae and infected root fragments have been shown to successfully colonize the roots of a host, germinating spores are considered to be the key players in new host establishment. Spores are commonly dispersed by fungal and plant burrowing herbivore partners, but some air dispersal capabilities are also known. Studies have shown that spore germination is specific to particular environmental conditions such as right amount of nutrients, temperature or host availability. It has also been observed that the rate of root system colonization is directly correlated to spore density in the soil. In addition, new data also suggests that AM fungi host plants also secrete chemical factors which attract and enhance the growth of developing spore hyphae towards the root system.

The necessary components for the colonization of Glomeromycota include, the host's fine root system, proper development of intracellular arbuscular structures, and a well-established external fungal mycelium. Colonization is accomplished by the interactions between germinating spore hyphae and the root hairs of the host or by development of appressoria between epidermal root cells. The process is regulated by specialized chemical signaling and by changes in gene expression of both the host and AM fungi. Intracellular hyphae extend up to the cortical cells of the root and penetrate the cell walls, but not the inner cellular membrane creating an internal invagination. The penetrating hyphae develop a highly branched structure called an arbuscule which have low functional periods before degradation and absorption by host's root cells. A fully developed arbuscular mycorrhizal structure facilitates the two-way movement of nutrients between the host and mutualistic fungal partner. The symbiotic association allows the host plant to respond better to environment stresses, and the non-photosynthetic fungi to obtain carbohydrates produced by photosynthesis.

Phylogeny

Initial studies of the Glomeromycota were based on the morphology of soil-borne sporocarps (spore clusters) found in or near colonized plant roots. Distinguishing features such as wall morphologies, size, shape, color, hyphal attachment and reaction to staining compounds allowed a phylogeny to be constructed. Superficial similarities led to the initial placement of genus *Glomus* in the unrelated family Endogonaceae. Following broader reviews that cleared up the sporocarp confusion, the Glomeromycota were first proposed in the genera *Acaulospora* and *Gigaspora* before being accorded their own order with the three families Glomaceae (now Glomeraceae), Acaulosporaceae and Gigasporaceae.

With the advent of molecular techniques this classification has undergone major revision. An analysis of small subunit (SSU) rRNA sequences indicated that they share a common ancestor with the Dikarya. Nowadays it is accepted that Glomeromycota consists of 4 orders:

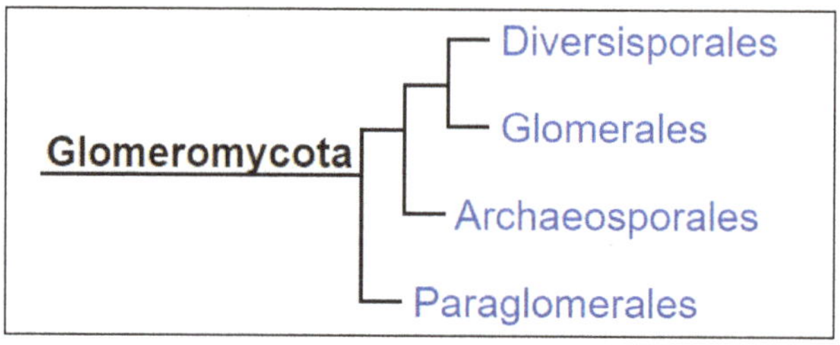

Several species which produce glomoid spores (i.e. spores similar to *Glomus*) in fact belong to other deeply divergent lineages and were placed in the orders, Paraglomerales and Archaeosporales. This new classification includes the Geosiphonaceae, which presently contains one fungus (*Geosiphon pyriformis*) that forms endosymbiotic associations with the cyanobacterium *Nostoc punctiforme* and produces spores typical to this division, in the Archaeosporales.

Work in this field is incomplete, and members of *Glomus* may be better suited to different genera or families.

Fossil Glomeromycota are known back 2200 million years as Diskagma, and also by Mesoproterozoic Horodyskia and Ordovician to Devonian Prototaxites.

Molecular Biology

The biochemical and genetic characterization of the Glomeromycota has been hindered by their biotrophic nature, which impedes laboratory culturing. This obstacle was eventually surpassed with the use of root cultures. The first mycorrhizal gene to be sequenced was the small-subunit ribosomal RNA (SSU rRNA). This gene is highly conserved and commonly used in phylogenetic studies so was isolated from spores of each taxonomic group before amplification through the polymerase chain reaction (PCR). A metatranscriptomic survey of the Sevilleta Arid Lands found that 5.4% of the fungal rRNA reads mapped to Glomeromycota. This result was inconsistent with previous PCR-based studies of community structure in the region; suggesting that previous PCR-based studies may have underestimated Glomeromycota abundance due to amplification biases.

CHYTRIDIOMYCOTA

Chytridiomycota are a division of zoosporic organisms in the kingdom Fungi, informally known as chytrids. Chytrids are one of the early diverging fungal lineages, and their membership in kingdom Fungi is demonstrated with chitin cell walls, a posterior whiplash flagellum, absorptive nutrition, use of glycogen as an energy storage compound, and synthesis of lysine by the α-amino adipic acid (AAA) pathway.

Chytrids are saprobic, degrading refractory materials such as chitin and keratin, and sometimes act as parasites. There has been a significant increase in the research of chytrids since the discovery of *Batrachochytrium dendrobatidis*, the causal agent of chytridiomycosis.

Classification

Species of Chytridiomycota have traditionally been delineated and classified based on development, morphology, substrate, and method of zoospore discharge. However, single spore isolates (or isogenic lines) display a great amount of variation in many of these features; thus, these features cannot be used to reliably classify or identify a species. Currently, taxonomy in Chytridiomycota is based on molecular data, zoospore ultrastructure and some aspects of thallus morphology and development.

In an older and more restricted sense (not used here), the term "chytrids" referred just to those fungi in the class Chytridiomycetes. Here, the term "chytrid" refers to all members of Chytridiomycota.

The chytrids have also been included among the Protoctista, but are now regularly classed as fungi.

In older classifications, chytrids, except the recently established order Spizellomycetales, were placed in the class Phycomycetes under the subphylum Myxomycophyta of the kingdom Fungi. Previously, they were placed in the Mastigomycotina as the class Chytridiomycetes. The other classes of the Mastigomycotina, the Hyphochytriomycetes and Oomycetes, were removed from the fungi to be classified as heterokont pseudofungi.

The class Chytridiomycetes has over 750 chytrid species distributed among ten orders. Additional classes include the Monoblepharidomycetes, with two orders, and the Hyaloraphidiomycetes with a single order.

Molecular phylogenetics, and other techniques such as ultrastructure analysis, has greatly increased the understanding of chytrid phylogeny, and led to the formation of several new zoosporic fungal phyla:

- The order Blastocladiales, originally within the Chytridiomycota, are now classified as a separate phylum, the Blastocladiomycota.
- The Neocallimastigales, originally an order of anaerobic fungi of the class Chytridiomycetes, found in the digestive tracts of herbivores, was later raised to a separate phylum, the Neocallimastigomycota.
- The Olpidiaceae, including the type genus *Olpidium*, formerly classified in the order Chytridiales, were raised to a separate phylum, the Olpidiomycota.

Life Cycle and Body Plan

Chytridiomycota are unusual among the Fungi in that they reproduce with zoospores. For most members of Chytridiomycetes, sexual reproduction is not known. Asexual reproduction occurs through the release of zoospores (presumably) derived through mitosis.

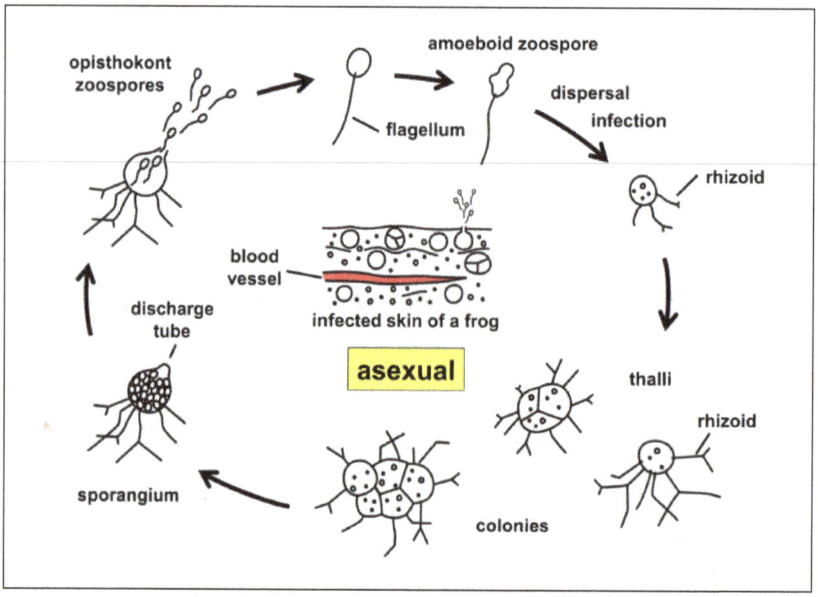

Life cycle of *Batrachochytrium dendrobatidis.*

Where it has been described, sexual reproduction of Chytridomycetes occurs via a variety of methods. It is generally accepted that the resulting zygote forms a resting spore, which functions as a means of surviving adverse conditions. In some members, sexual reproduction is achieved through the fusion of isogametes (gametes of the same size and shape). This group includes the notable plant pathogens *Synchytrium*. Some algal parasites practice oogamy: a motile male gamete attaches itself to a nonmotile structure containing the female gamete. In another group, two thalli produce tubes that fuse and allow the gametes to meet and fuse. In the last group, rhizoids of compatible strains meet and fuse. Both nuclei migrate out of the zoosporangium and into the conjoined rhizoids where they fuse. The resulting zygote germinates into a resting spore.

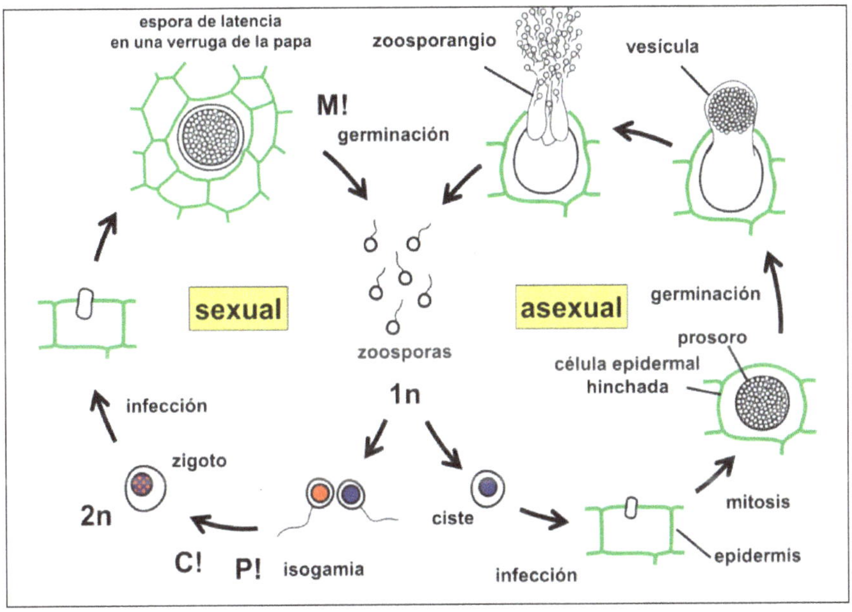

Life cycle of *Synchytrium endobioticum* in potato.

Sexual reproduction is common and well known among members of the Monoblepharidomycetes. Typically, these chytrids practice a version of oogamy: the male is motile and the female is stationary. This is the first occurrence of oogamy in kingdom Fungi. Briefly, the monoblephs form oogonia, which give rise to eggs, and antheridia, which give rise to male gametes. Once fertilized, the zygote either becomes an encysted or motile oospore, which ultimately becomes a resting spore that will later germinate and give rise to new zoosporangia.

Upon release from the germinated resting spore, zoospores seek out a suitable substrate for growth using chemotaxis or phototaxis. Some species encyst and germinate directly upon the substrate; others encyst and germinate a short distance away. Once germinated, enzymes released from the zoospore begin to break down the substrate and utilize it produce a new thallus. Thalli are coenocytic and usually form no true mycelium (having rhizoids instead).

Synchytrium endobioticum on potatoes.

Chytrids have several different growth patterns. Some are holocarpic, which means they only produce a zoosporangium and zoospores. Others are eucarpic, meaning they produce other structures, such as rhizoids, in addition to the zoosporangium and zoospores. Some chytrids are monocentric, meaning a single zoospore gives rise to a single zoosporangium. Others are polycentric, meaning one zoospore gives rise to many zoosporangium connected by a rhizomycelium. Rhizoids do not have nuclei while a rhizomycelium can.

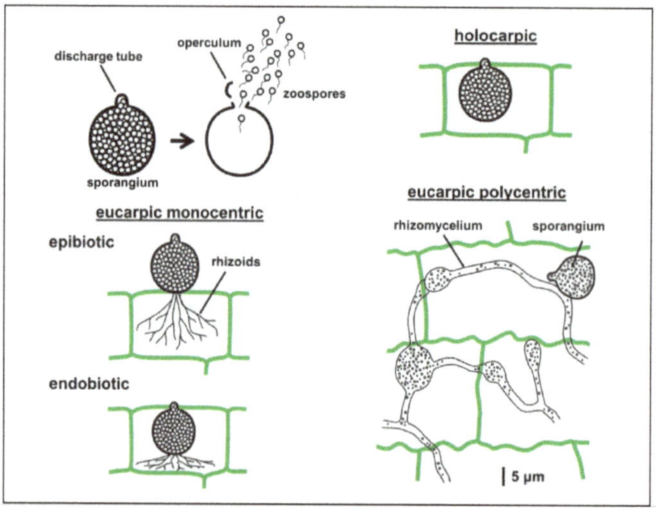

Types of chytrid thalli.

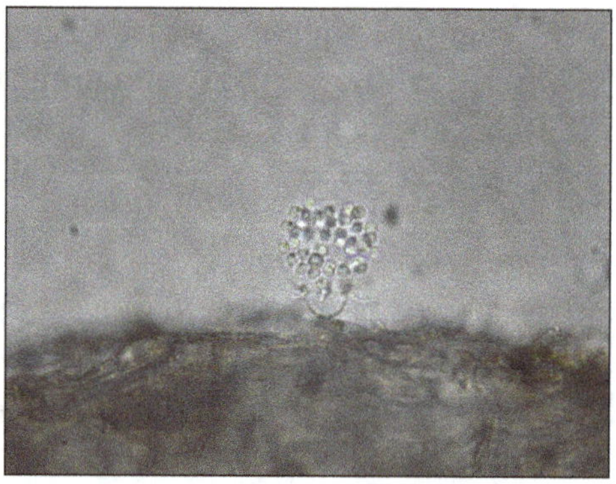

Zoospore discharge from *Phylctochytrium* sp.

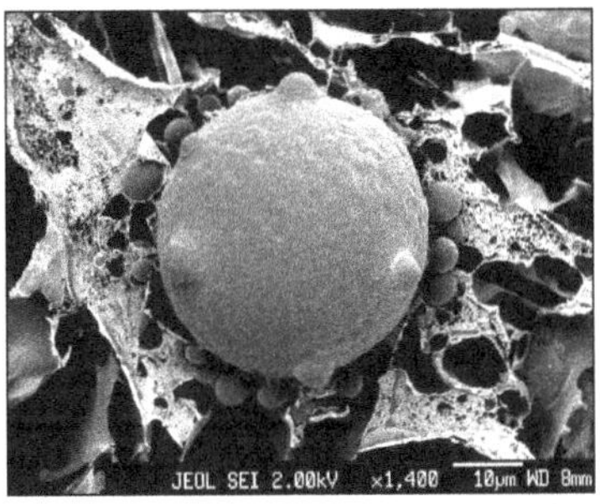

Sporangium and zoospores of the chytrid fungus *B. dendrobatidis*, under SEM.

Growth continues until a new batch of zoospores are ready for release. Chytrids have a diverse set of release mechanisms that can be grouped into the broad categories of operculate or inoperculate. Operculate discharge involves the complete or incomplete detachment of a lid-like structure, called an operculum, allowing the zoospores out of the sporangium. Inoperculate chytrids release their zoospores through pores, slits, or papillae.

Habitats

Chytrids are aquatic fungi, though those that thrive in the capillary network around soil particles are typically considered terrestrial. The zoospore is primarily a means of thoroughly exploring a small volume of water for a suitable substrate rather than a means of long-range dispersal.

Chytrids have been isolated from a variety of aquatic habitats, including peats, bogs, rivers, ponds, springs, and ditches, and terrestrial habitats, such as acidic soils, alkaline soils, temperate forest soils, rainforest soils, Arctic and Antarctic soils. This has led to the belief that many chytrid species are ubiquitous and cosmopolitan. However, recent taxonomic work has demonstrated that this ubiquitous and cosmopolitan morphospecies hide cryptic diversity at the genetic and ultrastructural levels. It was first thought aquatic chytrids (and other zoosporic fungi) were primarily active in fall, winter, and spring. However, recent molecular inventories of lakes during the summer indicate that chytrids are an active, diverse part of the eukaryotic microbial community.

One of the least expected terrestrial environments the chytrid thrive in are periglacial soils. The population of the Chytridiomycota species are able to be supported even though there is a lack of plant life in these frozen regions due to the large amounts of water in periglacial soil and pollen blowing up from below the timberline.

Ecological Functions

Chytrid parasites of marine diatoms. (A) Chytrid sporangia on Pleurosigma sp. The white arrow indicates the operculate discharge pore. (B) Rhizoids (white arrow) extending into diatom host. (C) Chlorophyll aggregates localized to infection sites (white arrows). (D and E) Single hosts bearing multiple zoosporangia at different stages of development. The white arrow in panel E highlights branching rhizoids. (F) Endobiotic chytrid-like sporangia within diatom frustule. Bars = 10 μm.

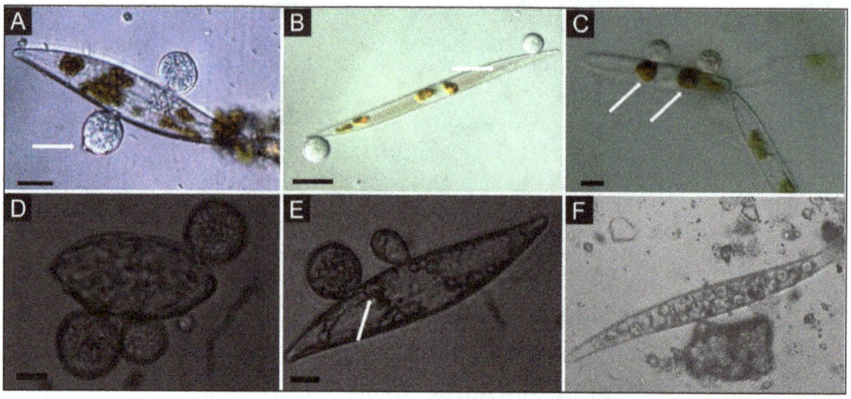

Dead frog with chytridiomycosis (*B. dendrobatidis*) signs.

Batrachochytrium Dendrobatidis

The chytrid *Batrachochytrium dendrobatidis* is responsible for chytridiomycosis, a disease of amphibians. Discovered in 1998 in Australia and Panama this disease is known to kill amphibians in large numbers, and has been suggested as a principal cause for the worldwide amphibian decline. Outbreaks of the fungus were found responsible for killing much of the Kihansi Spray Toad population in its native habitat of Tanzania, as well as the extinction of the golden toad in 1989. Chytridiomycosis has also been implicated in the presumed extinction of the Southern Gastric Brooding Frog, last seen in the wild in 1981, and the Northern Gastric Brooding Frog, last recorded in the wild in March 1985. The process leading to frog mortality is thought to be the loss of essential ions through pores made in the epidermal cells by the chytrid during its replication.

Other Parasites

Chytrids mainly infect algae and other eukaryotic and prokaryotic microbes. The infection can be so severe as to control primary production within the lake. It has been suggested that parasitic chytrids have a large effect on lake and pond food webs. Chytrids may also infect plant species; in particular, *Synchytrium endobioticum* is an important potato pathogen.

Saprobes

Arguably, the most important ecological function chytrids perform is decomposition. These ubiquitous and cosmopolitan organisms are responsible for decomposition of refractory materials, such as pollen, cellulose, chitin, and keratin. There are also chytrids that live and grow on pollen by attaching threadlike structures, called rhizoids, onto the pollen grains. This mostly occurs during asexual reproduction because the zoospores that become attached to the pollen continuously reproduce and form new chytrids that will attach to other pollen grains for nutrients. This colonization of pollen happens during the spring time when bodies of water accumulate pollen falling from trees and plants.

Fossil Record

The earliest fossils of chytrids are from the Scottish Rhynie chert, a Devonian-age lagerstätte with anatomical preservation of plants and fungi. Among the microfossils are chytrids preserved as parasites on rhyniophytes. These fossils closely resemble the modern genus *Allomyces*. Holocarpic chytrid remains were found in cherts from Combres in central France that date back to the late Visean. These remains were found along with eucarpic remains and are ambiguous in nature although they are thought to be of chytrids. Other chytrid-like fossils were found in cherts from the upper Pennsylvanian in the Saint-Etienne Basin in France, dating between 300 and 350ma.

MICROSPORIDIA

Microsporidia are obligate intracellular, spore-forming parasitic fungi.

Classification:

- Higher order taxa.
- Eukaryota; Fungi/Metazoa group; Fungi.

Species:

- Enterocytozoon bieneusi.
- Encephalitozoon cuniculi.
- Nosema spp.
- Pleistophora spp.

Description and Significance

Microsporidia (syn. microsporida) is a highly diverse phylum with which approximately

150 genera and over 1200 species are associated. These eukaryotic parasites infect mainly arthropods and fish, but have gained relevance in the medical community in the past few decades because of their increased infection of humans suffering from compromised immune symptoms, mainly associated with HIV. They also infect other mammals; the first known infection of a vertebrate was in a rabbit in 1922.

In arthropods, members of the genera Agmasoma, Amesoma, and Pleistophora infect and replace host tissues, resulting in eventual death and rendering the product unmarketable. They have a large negative impact on the fishing industries whose catches they may infect. Certain microsporidians are also extremely pathogenic in insects, and one, Nosema locustae, is marketed as a controllant against insects such as grasshoppers. Other species of Nosema, however, have a negative impact economically, since they infect beneficial insects such as honey bees (Nosema apis) and silkworms (Nosema bombycis). The clinical manifestations of the parasite in these varied hosts differs widely depending on the species, both of host and parasite.

At this time there are 13 microsporidians which are known to infect humans, primarily affecting those with immunocompromised disorders. The most prevalent pathogens of humans include Enterocytozoon bieneusi, whose clinical manifestations include diarrhea and pneumonia; Encephalitozoon cuniculi, whose symptoms include encephalitis and nephritis; Nosema sp., which affect the cornea; and Encephalitozoon intestinalis, which causes diarrhea and nephritis.

Genome Structure

Microsporidians have some of the smallest genomes known among eukaryotes; because of size constraints, their genomes do not grow with their genes. The genomes of thirteen microsporidians have been measured, ranging in size from 19.5 Mbp to 2.3 Mbp. Except for their small size, the genomes are characteristically similar to those of all eukaryotes. Once thought to be among the most primitive eukaryotic organisms, microsporidians are now known to possess some highly specialized organelles, utilized in the infestation process. These revelations have caused taxonomists to conclude that microsporidia are highly evolved fungi rather than ancient eukaryotes. Their reduced genomes and gene structures, the result of gene loss and compaction, reveal a highly evolved system rather than a primitive one. The eleven chromosomes of Encephalitozoon cuniculi's genome has been fully sequenced, allowing scientists to hypothesize that some of the microsporidia's "missing" organelles are instead simply derived from typical eukaryotic organelles.

Cell Structure and Metabolism

The most important and prevalent form taken during the microsporidian life cycle is that of a highly specialized single cell called the spore. When in this form the parasite is most easily recognizable, and individual species are identifiable. This is also the only form which is viable when the parasite is not infecting a living host. Spore morphology varies according to species, ranging in size from 1 - 40 microns, and taking on an ovoid, rod, spherical, or crescent shape. The spore features two extracellular walls surrounding a normal unit membrane, which contains the sporoplasm. This sporoplasm, which is the infectious material of the spore, contains one or two closely aligned nuclei, ribosome-enriched cytoplasm, and organelles used for infection. The most prominent among these organelles is the polar tube, located at the apex of the spore and affixed to the spore by way of an anchoring disk. From this anchoring disk the polar tube extends to the posterior end of the spore, where it comes into contact with another infection organelle, the posterior vacuole.

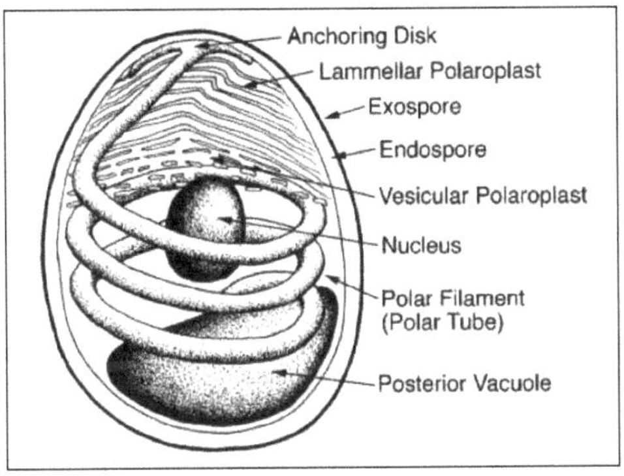

Unique spore structure of microsporidia.

Spore germination is initiated by a signal from the species' environment; the exact nature of these signals is not known for certain. Upon receiving this signal, the spore swells and the sporoplasmic membranes break down, culminating in the rupture of the anchoring disk. The polar tube is basically shot out of the spore, propelled by the internal pressure. After the polar tube is completely expelled, the sporoplasm enters the tube. There is enough force behind the polar tube to cause it to pierce the membrane of any nearby host cell, in which case the sporoplasm will enter and take over the other cell. The entire germination process takes as little as two seconds to be completed. Once inside the host cell, the parasite (referred to at this stage as the "meront") begins a series of developmental changes, which differ according to the species. More spores are produced through sporogony and eventually take over the host cell. In eukaryotic cells, energy is generated in the mitochondria; however, the cells of microsporidia reveal no such organelle. Microsporidia's exact mode of metabolism has yet to be fully understood, but they are obligate heterotrophs.

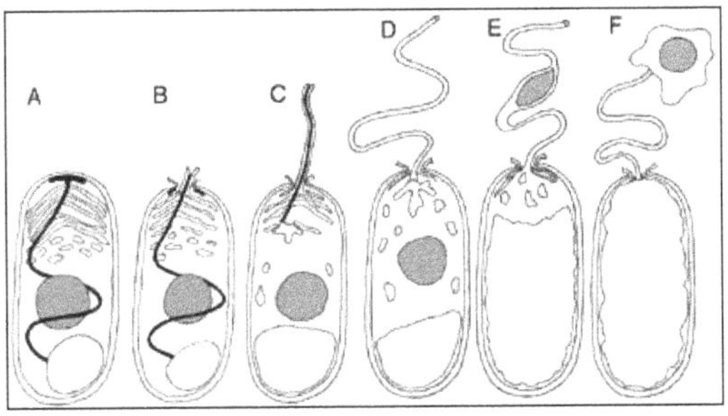

Stages of germination in microsporidia.

Ecology

Because the microsporidian species are so greatly numbered and their hosts so varied, these parasites can be found in many types of ecosystems. Researchers studying the difference between species which infected terrestrial hosts and those which infected aquatic hosts found that overwhelmingly terrestrial species could survive in low temperatures, while aquatic species could not. This is most likely because the terrestrial parasites need to survive cooler climates during the winter so that they can initiate infections in the early spring. Exposure to extremely dry conditions yielded similar results, with some species unable to survive. Similarly, the majority of microsporidians are unable to infect hosts when exposed to overly wet or humid conditions, possibly because an excessive amount of water causes the spores to germinate prematurely. Direct exposure to sunlight kills all species of microsporidia fairly quickly; some may be able to survive for a short amount of time, but most expire very quickly. Spores are spread from host to host by way of fecal matter, contaminated water and food, and inhalation. In humans, the most effective way to prevent the chronic diarrhea caused by microsporidia is cleanliness and decontamination of food. Currently human forms of these pathogens are most prevalent in countries or areas with high rates of AIDS among the population.

References

- Basidiomycota: newworldencyclopedia.org, Retrieved 14 March, 2020
- Lutzoni F; et al. (2004). "Assembling the fungal tree of life: progress, classification, and evolution of subcellular traits". American Journal of Botany. 91 (10): 1446–80. doi:10.3732/ajb.91.10.1446. PMID 21652303
- Hibbett, D.S.; et al. (March 2007). "A higher level phylogenetic classification of the Fungi". Mycological Research. 111 (5): 509–547. CiteSeerX 10.1.1.626.9582. doi:10.1016/j.mycres.2007.03.004. PMID 17572334
- James, T.Y.; et al. (2006). "A molecular phylogeny of the flagellated fungi (Chytridiomycota) and description of a new phylum (Blastocladiomycota)". Mycologia. 98 (6): 860–871. doi:10.3852/mycologia.98.6.860. PMID 17486963
- Schüßler, A.; et al. (December 2001). "A new fungal phylum, the Glomeromycota: phylogeny and evolution". Mycol. Res. 105 (12): 1413–1421. doi:10.1017/S0953756201005196
- Letcher, PM; et al. (2008). "Rhizophlyctidales--a new order in Chytridiomycota". Mycological Research. 112 (9): 1031–1048. doi:10.1016/j.mycres.2008.03.007. PMID 18701267
- Microsporidia: microbewiki.kenyon.edu, Retrieved 13 July, 2020

Types of Fungi

CHAPTER 4

There are different types of fungi in the world which includes mushroom, rust, puffball, polypore, morchella, truffle, yeast, mildews, molds, etc. The topics elaborated in this chapter will help in gaining a better perspective of the aspects related to these different types of fungi.

MUSHROOM

A mushroom is the fleshy, spore-bearing fruiting body (sporocarp or reproductive structure) of a fungus, typically produced above ground on soil or on its food source. The term "mushroom" also is used to designate the entire fungus with a fruiting body.

While the term mushroom is often associated with fungi that have a stem (stipe or stalk), a cap (pileus), and gills (lamella, the papery ribs under the cap of a mushroom), the term can refer to a wide variety of gilled fungi with or without stems and more generally any fruiting body.

The mushroom Amanita muscaria, commonly known as "fly agaric".

Mushrooms provide culinary, commercial, aesthetic, and ecological values. Since prehistoric times, people have consumed them. Edible varieties can be flavorful and provide B vitamins and minerals such as potassium, phosphorus, selenium, and iron; but mushrooms do not have many calories. Mushroom growing and sales have been an important industry, and the diverse forms and colors of mushrooms add to the wondrous nature for humans. Ecologically, various animals, such as rodents and birds, eat mushrooms, this makes mushrooms an important source of nutrition and energy in terrestrial food chains.

On the other hand, many species of mushrooms contain toxins that can cause human sickness or even death. Among the most deadly mushrooms are the destroying angel and the deathcap, members of the genus Amanita, which have killed many mushroom hunters.

Mushroom vs. Toadstool

The terms "mushroom" and "toadstool" date back centuries and were never precisely defined, nor was there consensus on its application. There is no scientific distinction between these terms. The term "toadstool" has often but not exclusively been applied to poisonous or inedible mushrooms, but has also been applied to those mushrooms that are edible and have the classic umbrella-like cap-and-stem form.

Between 1400 and 1600 C.E., the terms tadstoles, frogstooles, frogge stoles, tadstooles, tode stoles, toodys hatte, paddockstool, puddockstool, paddocstol, toadstoole, and paddockstooles were sometimes used synonymously with mushrom, mushrum, muscheron, mousheroms, mussheron, or musserouns. The term "mushroom" and its variations may have been derived from the French word mousseron in reference to moss (mousse). There may have been a direct connection to toads for toadstools.

However, there is no clear-cut delineation between edible and poisonous fungi, so that a "mushroom" technically may be edible, poisonous, or unpalatable. The term "toadstool" is currently used in storytelling when referring to poisonous or suspect mushrooms. The classic example of a toadstool is Amanita muscaria.

Growth Rates

Many species of mushrooms seemingly appear overnight, growing or expanding rapidly. This phenomenon is the source of several common expressions in the English language, including "to mushroom" or "mushrooming" (expanding rapidly in size or scope) and "to pop up like a mushroom" (to appear unexpectedly and quickly). In actuality, all species of mushrooms take several days to form primordial mushroom fruiting bodies, though they do expand rapidly by the absorption of fluids.

The cultivated mushroom as well as the common field mushroom initially form a minute fruiting body, referred to as the pin stage because of their small size. Slightly expanded, they are called buttons, once again because of the relative size and shape. Once such stages are formed, the mushroom can rapidly pull in water from its mycelium and expand, mainly by inflating preformed cells that took several days to form in the primordia.

Similarly, there are even more ephemeral mushrooms, like Parasola plicatilis (formerly Coprinus plicatlis), that literally appear overnight and may disappear by late afternoon on a hot day after rainfall. The primordia form at ground level in lawns in humid spaces under the thatch and after heavy rainfall or in dewy conditions balloon to full size in a few hours, release spores, and then collapse. They "mushroom" to full size.

He slang term "mushrooms" has served as a gang-related term for victims accidentally shot as collateral damage simply because they popped up suddenly, as do fungal mushrooms.

Not all mushrooms expand overnight; some grow very slowly and add tissue to their fruiting bodies by growing from the edges of the colony or by inserting hyphae.

Yellow, flower pot mushrooms (Leucocoprinus birnbaumii) at various states of development.

Identification

Identifying mushrooms requires a basic understanding of their macroscopic structure. Most are Basidiomycetes and gilled. Their spores, called basidiospores, are produced on the gills and fall in a fine rain of powder from under the caps as a result. At the microscopic level, the basidiospores are shot off of basidia and then fall between the gills in the dead air space. As a result, for most mushrooms, if the cap is cut off and placed gill-side-down overnight, a powdery impression reflecting the shape of the gills (or pores, or spines, etc.) is formed (when the fruiting body is sporulating). The color of the powdery print, called a spore print, is used to help classify mushrooms and can help to identify them. Spore print colors include white (most common), brown, black, purple-brown, pink, yellow, and cream, but almost never blue, green, or red.

While modern scientific identification of mushrooms is quickly becoming molecular, the standard methods for identification are still used by most and have developed into a fine art harking back to medieval times and the Victorian era. This ancient art is combined with microscopic examination. The presence of juices upon breaking, bruising reactions, odors, tastes, shades of color, habitat, habit, and season are all considered by mycologists, amateur and professional alike. Tasting and smelling mushrooms carries its own hazards because of poisons and allergens. Chemical spot tests are also used for some genera.

In general, identification to genus can often be accomplished in the field using a local mushroom guide. Identification to species, however, requires more effort; one must remember that a mushroom develops from a button stage into a mature structure, and only the latter can provide certain characteristics needed for the identification of the species. However, over-mature specimens lose features and cease producing spores. Many novices have mistaken humid water marks on paper for white spore prints, or discolored paper from oozing liquids on lamella edges for colored spore prints.

Classification

Two of the five major divisions (phyla) of fungi (Kingdom Fungi) are the Ascomycota and the Basidiomycota. The Ascomycota, commonly known as sac fungi or ascomycetes, form meiotic spores called ascospores, which are enclosed in a special sac-like structure called an ascus. This division includes morels, some mushrooms and truffles, as well as single-celled yeasts and many species that have only been observed undergoing asexual reproduction. Members of the Basidiomycota, commonly known as the club fungi or basidiomycetes, produce meiospores called basidiospores on club-like stalks called basidia. Most common mushrooms belong to this group, as well as rust and smut fungi, which are major pathogens of grains.

The genus Trichaptum, an example of a polypore, a mushroom without a stalk, fruiting on a log.

Typical mushrooms are the fruiting bodies of members of the order Agaricales, whose type genus is Agaricus and type species is the field mushroom, Agaricus campestris. The order Agaricales is placed in the class Agariomycetes of the phylum Basidiomycota. Agaricales is commonly known as the "gilled mushrooms".

However, in modern molecularly defined classifications, not all members of the order Agaricales have gills or produce mushroom fruiting bodies. Furthermore, many other gilled fungi, collectively called mushrooms, occur in other orders in the class Agaricomycetes. For example, chanterelles are in the Cantharellales, false chanterelles like Gomphus are in the Gomphales, milk mushrooms (Lactarius) and russulas (Russula) as well as

Lentinellus are in the Russulales, while the tough leathery genera Lentinus and Panus are among the Polyporales, but Neolentinus is in the Gloeophyllales, and the little pin-mushroom genus, Rickenella, along with similar genera, are in the Hymenochaetales.

Within the main body of mushrooms, in the order Agaricales, are common fungi like the common fairy-ring mushroom (Marasmius oreades), shiitake, enoki, oyster mushrooms, fly agarics, and other amanitas, magic mushrooms like species of Psilocybe, paddy straw mushrooms, shaggy manes, and so forth.

An atypical mushroom is the lobster mushroom (Hypomyces lactifluorum), which is an ascomycete that parasitizes other mushrooms, such as the Russula or Lactarius. This may result in a deformed, cooked-lobster-colored outside from the lobster mushroom and the white flesh of the parasitized basidiomycete muchroom on the inside.

Other mushrooms (those with fruiting bodies) are non-gilled. Some have pores underneath (and are usually called boletes), others have spines, such as the hedgehog mushroom and other tooth fungi, and so on. "Mushroom" has been used for polypores, puffballs, jelly fungi, coral fungi, bracket fungi, stinkhorns, and cup fungi.

Thus the term mushroom has more than one common application to macroscopic fungal fruiting bodies and associated organisms than one having precise taxonomic meaning. According to Chang and Miles, there are approximately fourteen thousand described species of mushrooms.

Human Use

Edible Mushrooms

The button mushroom (Agaricus bisporus), one of the most widely cultivated mushrooms in the world.

Edible mushrooms are used extensively in cooking, in many cuisines (notably Chinese, European, and Japanese). Though mushrooms are commonly thought to have little nutritional value, many species are high in fiber and provide vitamins such as thiamine (B_1), riboflavin (B_2), niacin (B_3), biotin (B_7), cobalamins (B_{12}), and ascorbic acid (C), as well as minerals, including iron, selenium, potassium, and phosphorus. Mushrooms have been gaining a higher profile for containing the antioxidants ergothioneine and selenium.

Most mushrooms that are sold in supermarkets have been commercially grown on mushroom farms. The most popular of these, Agaricus bisporus, is safe for most people to eat because it is grown in controlled, sterilized environments, though some individuals do not tolerate it well. Several varieties of A. bisporus are grown commercially, including whites, crimini, and portabello. Other cultivated species now available at many grocers include shiitake, maitake or hen-of-the-woods, oyster, and enoki.

People who collect mushrooms for consumption are known as mycophagists, and the act of collecting them for such is known as mushroom hunting, or simply "mushrooming".

Toxic Mushrooms

Of central interest is the chemical properties of mushrooms and the fact that many species produce secondary metabolites that render them toxic, mind-altering, or even bioluminescent.

The Panther cap (Amanita pantherina), a toxic mushroom.

There are a number of species of mushroom that are poisonous, and although some resemble certain edible species, eating them could be fatal. Eating mushrooms gathered in the wild is risky and should not be undertaken by individuals unknowledgeable in mushroom identification, unless the individuals limit themselves to a small number of good edible species that are visually distinctive. More generally, and particularly with gilled mushrooms, separating edible from poisonous species requires meticulous attention to detail; there is no single trait by which all toxic mushrooms can be identified, nor one by which all edible mushrooms can be identified.

Though there are only a small number of deadly species, several have particularly severe and unpleasant symptoms. Toxicity likely plays a role in protecting the function of the basidiocarp: The mycelium has expended considerable energy and protoplasmic material to develop a structure to efficiently distribute its spores. One defense against consumption and premature destruction is the evolution of chemicals that render the mushroom inedible, either causing the consumer to vomit the meal or avoid consumption altogether.

Psychoactive Mushrooms

Psilocybin mushrooms possess psychedelic properties. They are commonly known as "magic mushrooms" or "shrooms" and are available in many parts of the world, though a number of countries have outlawed their sale. Some mushrooms are eaten for their psychoactive effects, such as fly agaric, which is used for shamanic purposes by tribes in northeast Siberia. They have also been used in the West to enhance religious experiences. Because of their psychoactive properties, some mushrooms have played a role in native medicine, where they have been used to effect mental and physical healing, and to facilitate visionary states. One such ritual is the Velada ceremony. A representative figure of traditional mushroom use is the shaman and curandera (priest-healer) María Sabina.

As with many psychoactive substances, the effects of psychedelic mushrooms are subjective and unpredictable, with the effect ranging from exhilarating to distressing.

Medicinal Mushrooms

Currently, many species of mushrooms and fungi used in thousand-year-old folk medicine practices are under intense study by ethnobotanists and medical researchers. Maitake, shiitake, chaga, and reishi are prominent among those being researched for their potential anti-cancer, anti-viral, or immunity-enhancing properties. Psilocybin, originally an extract of certain psychedelic mushrooms, is being studied for its ability to help people suffering from mental disease, such as obsessive-compulsive disorder. Minute amounts have been reported to stop cluster and migraine headaches.

Other Uses

Mushrooms can be used for dyeing wool and other natural fibers. The chromophores of mushrooms are organic compounds that produce strong and vivid colors, and all colors of the spectrum can be achieved with mushroom dyes. Before the invention of synthetic dyes, mushrooms were the primary source of textile dyes. This technique has survived in Finland, and many Middle Ages re-enactors have revived the skill.

Some fungi, types of polypores loosely called mushrooms, have been used as fire starters (known as tinder fungi). Ötzi the Iceman (a well-preserved natural mummy of a man from about 3300 B.C.E.) was found carrying such fungi. Mushrooms and other fungi will likely play an increasingly important role in the development of effective biological remediation and filtration technologies.

RUST

Rusts are plant diseases caused by pathogenic fungi of the order Pucciniales.

An estimated 168 rust genera and approximately 7,000 species, more than half of which belong to the genus Puccinia, are currently accepted. Rust fungi are highly specialized plant pathogens with several unique features. Taken as a group, rust fungi are diverse and affect many kinds of plants. However, each species has a very narrow range of hosts and cannot be transmitted to non-host plants. In addition, most rust fungi cannot be grown easily in pure culture.

A single species of rust fungi may be able to infect two different plant hosts in different stages of its life cycle, and may produce up to five morphologically and cytologically distinct spore-producing structures viz., spermogonia, aecia, uredinia, telia and basidia in successive stages of reproduction. Each spore type is very host specific, and can typically infect only one kind of plant.

Rust fungi are obligate plant pathogens that only infect living plants. Infections begin when a spore lands on the plant surface, germinates, and invades its host. Infection is limited to plant parts such as leaves, petioles, tender shoots, stem, fruits, etc. Plants with severe rust infection may appear stunted, chlorotic (yellowed), or may display signs of infection such as rust fruiting bodies. Rust fungi grow intracellularly, and make spore-producing fruiting bodies within or, more often, on the surfaces of affected plant parts. Some rust species form perennial systemic infections that may cause plant deformities such as growth retardation, witch's broom, stem canker, galls, or hypertrophy of affected plant parts.

Rusts get their name because they are most commonly observed as deposits of powdery rust-coloured or brown spores on plant surfaces. The Roman agricultural festival Robigalia has ancient origins in combating wheat rust.

Impacts

Rusts are considered among the most harmful pathogens to agriculture, horticulture and forestry. Rust fungi are major concerns and limiting factors for successful cultivation of agricultural and forest crops. White pine blister rust, wheat stem rust, soybean rust, and coffee rust are examples of notoriously damaging, economically important crops.

Life Cycle

All rusts are obligate parasites, meaning that they require a living host to complete their life cycle. They generally do not kill the host plant but can severely reduce growth and yield. Cereal crops can be devastated in one season; oak trees infected in the main stem within their first five years by the rust Cronartium quercuum often die.

Rust fungi can produce up to five spore types from corresponding fruiting body types during their life cycle, depending on the species. Roman numerals have traditionally been used to refer to these morphological types.

- 0-Pycniospores (Spermatia) from Pycnidia: These serve mainly as haploid gametes in heterothallic rusts.

- I-Aeciospores from Aecia: These serve mainly as non-repeating, dikaryotic, asexual spores, and go on to infect the primary host.
- II-Urediniospores from Uredia (Uredinia): These serve as repeating dikaryotic vegetative spores. These spores are referred to as the repeating stage because they can cause auto-infection on the primary host, re-infecting the same host on which the spores were produced. They are often profuse, red/orange, and a prominent sign of rust disease.
- III-Teliospores from Telia: These dikaryotic spores are often the survival/ overwintering stage of the life cycle. They usually do not infect a plant directly; instead they germinate to produce basidia and basidiospores.
- IV-Basidiospores from Teliospores: These windborne haploid spores often infect the alternate host in Spring. They are rarely observed outside of the laboratory.

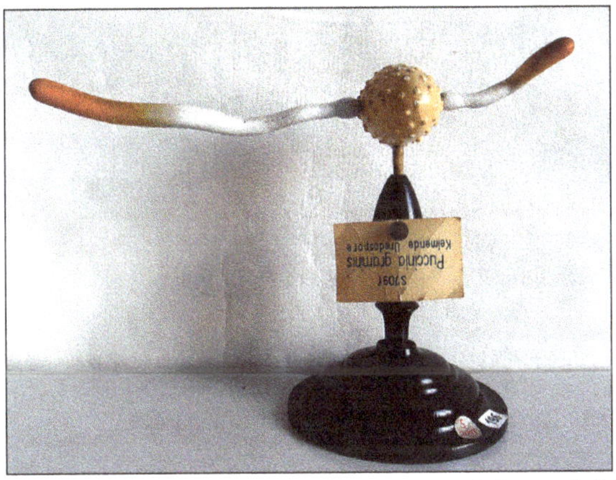

Germinating urediniospore of Puccinia graminis, model from the late 19th century, Botanical Museum Greifswald.

Rust fungi are often categorized by their life cycle. Three basic types of life cycles are recognized based on the number of spore types as macrocyclic, demicyclic, and microcyclic. The macrocyclic life cycle has all spore states, the demicyclic lacks the uredinial state, and the microcyclic cycle lacks the basidial, pycnial, and the aecial states, thus possess only uredinia and telia. Spermagonia may be absent from each type but especially the microcyclic life cycle. In macrocyclic and demicyclic life cycles, the rust may be either host alternating (heteroecious) (i.e., the aecial state is on one kind of plant but the telial state on a different and unrelated plant), or non-host alternating (autoecious) (i.e., the aecial and telial states on the same plant host). Heteroecious rust fungi require two unrelated hosts to complete their life cycle, with the primary host being infected by aeciospores and the alternate host being infected by basidiospores. This can be contrasted with an autoecious fungus which can complete all parts of its life cycle on a single host species. Understanding the life cycles of rust fungi allows for proper disease management.

Host Plant-rust Fungus Relationship

There are definite patterns of relationship with host plant groups and the rust fungi that parasitize them. Some genera of rust fungi, especially Puccinia and Uromyces, comprise species that are capable of parasitizing plants of many families. Other rust genera appear to be restricted to certain plant groups. Host restriction may, in heteroecious species, apply to both phases of life cycle or to only one phase.

Infection Process

The spores of rust fungi may be dispersed by wind, water or insect vectors. When a spore encounters a susceptible plant, it can germinate and infect plant tissues. A rust spores typically germinates on a plant surface, growing a short hypha called a germ tube. This germ tube may locate a stoma by a touch responsive process known as thigmotropism. This involves orienting to ridges created by epidermal cells on the leaf surface, and growing directionally until it encounters a stoma.

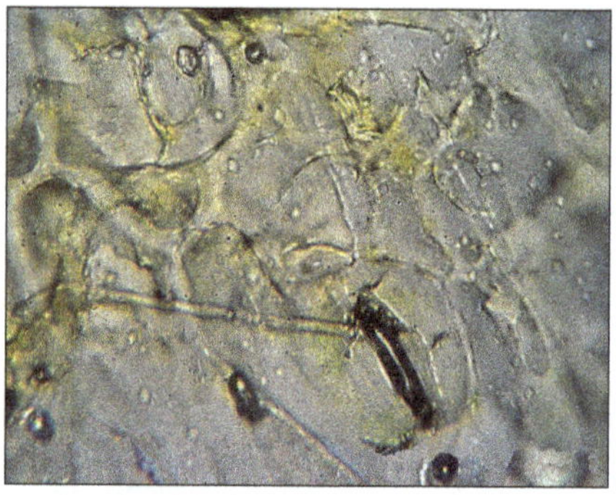

Rust hypha attacking stoma.

Over the stoma, a hyphal tip produces an infection structure called an appressorium. From the underside of an appressorium, a slender hypha grows downward to infect plant cells. It is thought that the whole process is mediated by stretch-sensitive calcium ion channels located in the tip of the hypha, which produce electric currents and alter gene expression, inducing appressorium formation.

Once the fungus has invaded the plant, it grows into plant mesophyll cells, producing specialized hyphae known as haustoria. The haustoria penetrate cell walls but not cell membrances: plant cell membranes invaginate around the main haustorial body forming a space known as the extra-haustorial matrix. An iron and phosphorus rich neck band bridges the plant and fungal membranes in the space between the cells for water flow, known as the apoplast, thus preventing the nutrients reaching the plant's cells. The haustorium contains amino acid and hexose sugar transporters and H^+-ATPases which are used for active transport of nutrients from the plant, nourishing the fungus. The fungus continues growing, penetrating more and more plant cells, until spore growth occurs. The process repeats every 10 – 14 days, producing numerous spores that can be spread to other parts of the same plant, or to new hosts.

Management of Rust Fungi Diseases

The control methods of rust fungus diseases depend largely on the life cycle of the particular pathogen. The following are examples of disease management plans used to control macrocyclic and demicyclic diseases:

Macrocyclic Disease: Developing a management plan for this type of disease depends largely on whether the repeating stages (urediniospores) occur on the economically important host plant or the alternate host. For example, the repeating stage in white pine blister rust disease does not occur on white pines but on the alternate host, Ribes spp. During August and September Ribes spp. give rise to teliospores which infect white pines. Removal of the alternate host disrupts the life cycle of the rust fungi Cronartium ribicola, preventing the formation of basidiospores which infect the primary host. Although spores from white pines cannot infect other white pines, survival spores may overwinter on infected pines and reinfect Ribes spp. the following season. Infected tissue is removed from white pines and strict quarantines of Ribes spp. are maintained in high risk areas.

Puccinia graminis is a macrocyclic heteroecious fungus that causes wheat stem rust disease. The repeating stage in this fungus occurs on wheat and not the alternate host, barberry. The repeating stage allows the disease to persist in wheat even though the alternate host may be removed. Planting resistant crops is the ideal form of disease prevention, however, mutations can give rise to new strains of fungi that can overcome plant resistance. Although the disease cannot be stopped by removal of the alternate host, the life cycle is disrupted and the rate of mutation is decreased because of reduced genetic recombination. This allows resistance bred crops to remain effective for a longer period of time.

Demicyclic Disease: Because there is no repeating stage in the life cycle of demicyclic fungi, removal of the primary or the alternate host will disrupt the disease cycle. This method, however, is not highly effective in managing all demicyclic diseases. Cedar-apple rust disease, for example, can persist despite removal of one of the hosts since spores can be disseminated from long distances. The severity of Cedar-apple rust disease can be managed by removal of basidiospore producing galls from junipers or the application of protective fungicides to junipers.

Home Control

Rust diseases are very hard to treat. Fungicides such as Mancozeb or Triforine may help but may never eradicate the disease. Some organic preventative solutions are available and sulphur powder is known to stop spore germination. High standards of hygiene, good soil drainage, and careful watering may minimize problems. Any appearance of rust must be immediately dealt with by removing and burning all affected leaves. Composting, or leaving infected vegetation on the ground will spread the disease.

Commercial Control

In some large acreage crops, fungicides are applied by air. The process is expensive and fungicide application is best reserved for seasons when foliar diseases are severe. Research indicates, the higher the foliar disease severity, the greater the return from the use of fungicides. Southern corn rust disease, can be confused with common rust. Southern rust's distinguishing characteristic is that pustules form mostly on the upper leaf surface and spores are more orange in color. Southern rust spreads more quickly and has a higher economic impact when hot, humid weather conditions persist. Timely fungicide applications to control southern rust are more crucial than with common rust.

A variety of preventative methods can be employed for rust diseases:

- High moisture levels may exacerbate rust disease symptoms. The avoidance of overhead watering at night, using drip irrigation, reducing crop density, and using fans to circulate air flow may decrease disease severity.
- The use of rust resistant plant varieties.
- Crop rotation can break the disease cycle because many rusts are host specific and do not persist long without their host.
- Inspection of imported plants and cuttings for symptoms. It is important to continuously observe the plants because rust diseases have a latent period (plant has the disease but shows no symptoms).
- Use of disease-free seed can reduce incidence for some rusts.

PUFFBALL

Puffballs are fungi, so named because clouds of brown dust-like spores are emitted when the mature fruitbody bursts or is impacted. Puffballs are in the division Basidiomycota and encompass several genera, including Calvatia, Calbovista and Lycoperdon. True puffballs do not have a visible stalk or stem. The puffballs were previously treated as a taxonomic group called the Gasteromycetes or Gasteromycetidae, but they are now known to be a polyphyletic assemblage.

The distinguishing feature of all puffballs is that they do not have an open cap with spore-bearing gills. Instead, spores are produced internally, in a spheroidal fruitbody called a gasterothecium (gasteroid ('stomach-like') basidiocarp). As the spores mature, they form a mass called a gleba in the centre of the fruitbody that is often of a distinctive color and texture. The basidiocarp remains closed until after the spores have been released from the basidia. Eventually, it develops an aperture, or dries, becomes brittle, and splits, and the spores escape. The spores of puffballs are statismospores rather than ballistospores, meaning they are not forcibly extruded from the basidium. Puffballs and similar forms are thought to have evolved convergently (that is, in numerous independent events) from Hymenomycetes by gasteromycetation, through secotioid stages. Thus, 'Gasteromycetes' and 'Gasteromycetidae' are now considered to be descriptive, morphological terms (more properly gasteroid or gasteromycetes, to avoid taxonomic implications) but not valid cladistic terms.

Lycoperdon perlatum.

Calbovista subsculpta.

Stalked puffballs do have a stalk that supports the gleba. None of the stalked puffballs are edible as they are tough and woody mushrooms. The Hymenogastrales and Enteridium lycoperdon, a slime mold, are the false puffballs. A gleba which is powdery on maturity is a feature of true puffballs, stalked puffballs and earthstars. False puffballs are hard like rock or brittle. All false puffballs are inedible, as they are tough and bitter to taste. The genus Scleroderma, which has a young purple gleba, should also be avoided.

Calvatia cyathiformis: Puffballs are found in several genera of the division Basidiomycota.

Puffballs were traditionally used in Tibet for making ink by burning them, grinding the ash, then putting them in water and adding glue liquid and "a nye shing ma decoction", which, when pressed for a long time, made a black dark substance that was used as ink.

Edibility and Identification

While most puffballs are not poisonous, some often look similar to young agarics, and especially the deadly Amanitas, such as the death cap or destroying angel mushrooms.

It is for this reason that all puffballs gathered in mushroom hunting are cut in half lengthwise. Young puffballs in the edible stage, before maturation of the gleba, have undifferentiated white flesh within; whereas the gills of immature Amanita mushrooms can be seen if they are closely examined.

Spores coming out of puffball fungus.

The giant puffball, Calvatia gigantea (earlier classified as Lycoperdon giganteum), reaches a foot (30 cm) or more in diameter, and is difficult to mistake for any other fungus. It has been estimated that a large specimen of this fungus when mature will produce around 7×10^{12} spores. If collected before spores have formed, while the flesh is still white, it may be cooked as slices fried in butter, with a strong earthy, mushroom flavor.

Not all true puffball mushrooms are without stalks. Some may also be stalked like the Podaxis pistillaris which is also called the "false shaggy mane".

Puffball mushrooms on sale at a market in England.

POLYPORE

Polypores are a group of fungi that form large fruiting bodies with pores or tubes on the underside. They are a morphological group of basidiomycetes - like gilled mushrooms and hydnoid fungi, and not all polypores are closely related to each other. Polypores are also called bracket fungi, and their woody fruiting bodies are called conks.

Polypores (Ganoderma sp.) growing on a tree in Borneo.

Most polypores inhabit tree trunks or branches consuming the wood, but some soil-inhabiting species form mycorrhiza with trees. Polypores and their relatives corticioid fungi are the most important agents of wood decay. Thus, they play a very significant role in nutrient cycling and carbon dioxide production of forest ecosystems.

Over one thousand polypore species have been described to science, but a large part of the diversity is still unknown even in relatively well-studied temperate areas. Polypores are much more diverse in old natural forests with abundant dead wood than in younger managed forests or plantations. Consequently, a number of species have declined drastically and are under threat of extinction due to logging and deforestation.

Trametes versicolor, a colorful bracket fungus, commonly known as turkey tail.

Polypores are used in traditional medicine, and they are actively studied for their medicinal value and various industrial applications. Several polypore species are serious pathogens of plantation trees and are major causes of timber spoilage.

A bracket fungus (Pycnoporus sp.) with a tough, woody cap.

Bracket fungi, or shelf fungi, are among the many groups of fungi that compose the division Basidiomycota. Characteristically, they produce shelf- or bracket-shaped or occasionally circular fruiting bodies called conks that lie in a close planar grouping of separate or interconnected horizontal rows. Brackets can range from only a single row of a few caps, to dozens of rows of caps that can weigh several hundred pounds. They are mainly found on trees (living and dead) and coarse woody debris, and may resemble mushrooms. Some form annual fruiting bodies while others are perennial and grow larger year after year. Bracket fungi are typically tough and sturdy and produce their spores, called basidiospores, within the pores that typically make up the undersurface.

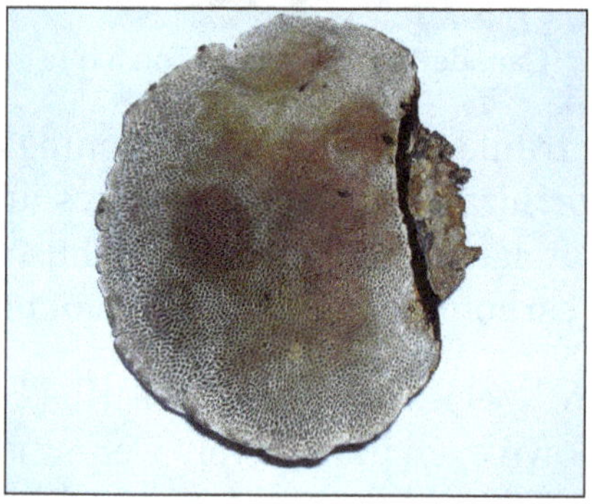

The blushing bracket showing the red bruising, which is one identification characteristic.

Classification

Because bracket fungi are defined by their growth form rather than phylogeny, the group contains members of multiple clades. Although the term classically was reserved for polypores, molecular studies have revealed some odd relationships. The beefsteak fungus, a well-known bracket fungus, is actually a member of the agarics. Other examples of bracket fungi include the sulphur shelf, birch bracket, dryad's saddle, artist's conk, and turkey tail. The name polypores is often used for a group that includes many of the hard or leathery fungi, which often lack a stipe, growing straight out of wood.

The group includes many different shapes and forms that are common in the tropical forests, including the hard 'cup fungi' and the 'shell', 'plate' and 'bracket' fungus commonly found growing off logs and still standing dead trees.

Laetiporus sulphureus.

The fungal individual that develops the fruit bodies that are identified as polypores resides in soil or wood as mycelium. Polypores are often restricted to either deciduous (angiosperm) or conifer (gymnosperm) host trees. Some species depend on a single tree genus (e.g. Piptoporus betulinus on birch, Perenniporia corticola on dipterocarps).

Forms of polypore fruit bodies range from mushroom-shaped to thin effused patches (crusts) that develop on dead wood. Perennial fruit bodies of some species growing on living trees can grow over 80 years old (e.g. Phellinus igniarius). Most species of polypores develop new, short-lived fruit bodies annually or several times every year. Abundant fruit takes place during the autumn or rainy season.

Structure of the fruit bodies is simple. Effused or resupinate fruit bodies typically consist of two layers - a tube layer of vertically arranged tubes that open downwards, and supporting layer called subiculum that supports and attached the tubes to substrate. In fruit bodies with a cap (pileate fruit bodies) the tissue between upper surface and the pore layer is called context. A few polypores (e.g. Fomes fomentarius and Inocutis rhaedes) also have a core between context and substrate. A minority of polypores also have a stalk (stipe) that attach to the cap either laterally or centrally depending on the species.

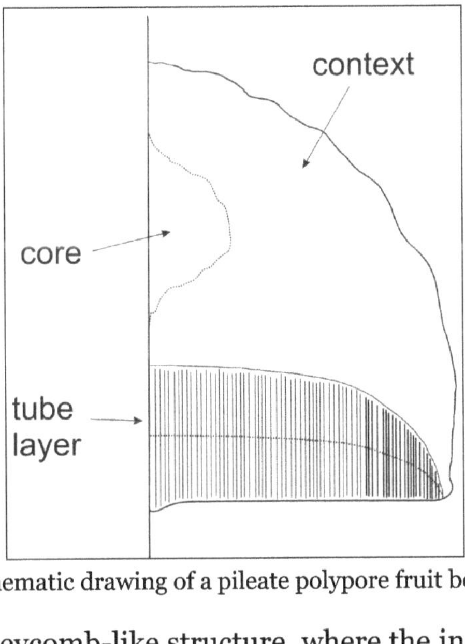

Schematic drawing of a pileate polypore fruit body.

Polypore tubes are a honeycomb-like structure, where the individual tubes have fused together. Their sides are covered with a spore-forming surface, the hymenium. The tubes offer shelter for developing spores and help to increase the area of the spore-producing surface. Pore size and shape vary a lot between species, but little within a species – some Hexagonia spp. have 5 mm wide pores whereas pores of Antrodiella spp. are invisible to naked eye with 15 pores per mm. Generally the larger the pores, the larger the spores. A few polypores produce asexual spores (chlamydospores or conidia) in the upper surface of their cap (e.g. Echinopora aculeifera, Oligoporus ptychogaster) or without the presence of a sexual fruit body (e.g. Inonotus rickii, Heterobasidion spp.).

Polyporus sp. fruit bodies with stalk (Indonesia).

Perennial conk of Fomitopsis pinicola on spruce.

Ecology

Bracket fungi often grow in semi-circular shapes, looking like trees or wood. They can be parasitic, saprotrophic, or both. One of the more common genera, *Ganoderma*, can grow large thick shelves that may contribute to the death of the tree, and then feed off the wood for years after. Their hardiness means they are very resilient and can live for quite a long time, with many species even developing beautiful multi-coloured circles of colour that are actually annual growth rings. Polypores are among the most efficient decomposers of lignin and cellulose, the main components of wood. Due to this ability they dominate communities of wood-rotting organisms in land ecosystems along with corticioid fungi. Through decomposing tree trunks they recycle a major part of nutrients in forests.

Only basidiomycetes are known to initiate lignin degradation (i.e. cause white rot). A recent study linked the end of formation of large-scale coal deposits in the end of the Carboniferous period 300 million years ago to the evolution of ligning-degrading basidiomycetes. More efficient degradation of wood by fungi meant less plant material (coal-to-be) accumulating in the soil.

On the other hand, most brown-rot fungi are polypores. These species have lost their lignin degradation ability but are very efficient in degrading cellulose. Brown-rot fungi are prevalent on conifer hosts and open, sun-exposed habitats. The fungal community in any single trunk may include both white-rot and brown-rot species, complementing each other's wood degradation strategies.

Polypores and other decomposer fungi are the first step in food chains that feed on decomposed plant material. A rich fauna of insects, mites and other invertebrates feed on polypore mycelium and fruiting bodies, further providing food for birds and other larger animals. Woodpeckers and other hole nesting birds typically carve their nests in softer wood decomposed by polypores.

Threats

Almost all polypores are dependent on trees for their survival. Deforestation and intensive forest management cause declines in polypore abundance and diversity. For many species the changes can be too much, and they start a slow slide towards extinction. Since most polypore species are relatively widespread, this process is typically slow. Regional extinctions can happen relatively quickly and have been documented (for instance Antrodia crassa in North Europe).

Polypores can decline for many reasons. They can be dependent on a single host or a very special habitat. For instance Echinodontium ballouii has been found only in Atlantic white cedar swamps in the Northeastern USA. Species can be dependent on very old tree individuals like Bridgeoporus nobilissimus of the Northwestern USA. Both of these species also have a rather restricted range, making them more vulnerable to extinction.

In addition to the host tree individual, the characteristics of the surrounding habitat also matter. Some species prefer closed-canopy forest with a moist, even microclimate that could be disturbed for instance by logging (e.g. Skeletocutis jelicii). Others suffer from lack of open forest-fire habitat in areas where fire suppression is done (e.g. Gloeophyllum carbonarium in Nordic countries where forest fires are part of the natural forest dynamics).

For most declining species the main problem is lack of dead wood in the forest. When suitable tree trunks are too sparse in the landscape, not all species are able to spread to new trunks after old ones have been consumed, causing the population to decline and eventually vanish. Thus, species that are abundant in old-growth forests with abundant dead wood can be totally absent from managed forests. For instance Amylocystis lapponica and Fomitopsis rosea are dominant species in North European old-growth spruce forests from Poland to Norway, but absent in managed forests.

Climate change may cause a problem for polypores that are already dependent on a few fragments of old-growth forests and may be unable to migrate with changing vegetation.

Indicator Value

Polypores have been used as indicator species of healthy natural forests or old-growth forests in Europe. They are good indicators of invertebrate diversity on dead wood and include many endangered species. Polypores make good indicators because they are relatively easy to find – many species produce conspicuous and long-lasting fruiting bodies – and because they can be identified in the field.

The first indicator list of polypores widely used in forest inventories and conservation work was developed in northern Sweden in 1992 ("Steget före" method). "Steget före" list included six polypores in three value classes. In Finland, a list of 30 species for spruce-dominated forests was published in 1993 and widely adopted. Later a similar list for pine-dominated forests was published. Longer lists of indicator species have since been published in Sweden.

Many indicator species are red-listed, but not necessarily all. National red lists of fungi typically include many polypores and are used as indicator lists of conservation value in many European countries.

Classification

For most of 20th century polypores were treated as a family, the Polyporaceae. Reconstructions of family tree of fungi show that the poroid fruiting body has evolved numerous times in the past. Modern DNA-based evolutionary classification places polypores to at least 12 orders. The orders containing most polypore species are the Polyporales (genera such as Fomes, Polyporus and Trametes) and Hymenochaetales (e.g. Oxyporus, Phellinus and Trichaptum). Economically perhaps the most significant polypores Heterobasidion spp., pests of conifer plantations, belong to the Russulales. Other polypore orders are the Agaricales, Amylocorticiales, Auriculariales, Boletales, Cantharellales, Gloeophyllales, Sebacinales, Thelephorales and Trechisporales.

The Polyporales in the modern sense are not only polypores but also other fruiting body types such as crust fungi, hydnoid fungi and agaricoid mushrooms. The term polypore describing a morphological group should not be confused with the taxonomic groups Polyporales or Polyporaceae of the modern literature.

Currently polypores are divided into about 170 genera. That number is bound to rise significantly through better understanding of evolutionary relationships between species and through mapping of uncovered diversity in the tropics. All in all classification of polypores is in flux.

Delimitation (Morphology)

Irpex lacteus with irregular pores.

Most polypores have a poroid hymenium, but not all species. A few, for instance Elmerina holophaea and Lenzites betulina, form gills like agarics but are still considered polypores, since in all other respects they are similar to closely related polypores, forming tough fruiting bodies on the wood. A couple of species where the tubes have not fused together in a honey-comb manner are variably classified as polypores or not (e.g. Porotheleum fimbriatum). There's no clear distinction between polypores and hydnoid fungi - some polypores with irregularly poroid lower surface have been considered both polypores and hydnoid fungi (e.g. Echinodontium tinctorium, Irpex lacteus).

Bolete mushrooms are a separate morphological group not included in polypores even though they have tubes. Fleshy fruiting bodies with a stalk and microscopic characters separate boletes from polypores.

Elmerina holophaea, a polypore with gills.

Uses

Some species of bracket fungi are edible, such as sulphur polypore; the lingzhi mushroom is another, which is used in Chinese medicine. They can also be used as a wick in an oil/fat lamp.

The tinder fungus (Fomes fomentarius) has been used as tinder since at least the time of Ötzi the Iceman. It has also been used to make a material similar to leather.

Ganoderma applanatum, the artist's conk, is used as a substrate for drawings. Fresh specimens develop dark brown lines when drawn upon with a stylus. The lines become permanent when the specimen is dried.

Medicinal Uses

Most polypores are edible or at least non-toxic, however one genus of polypores has members that are poisonous. Polypores from the genus Hapalopilus have caused poisoning in several people with effects including kidney dysfunction and deregulation of central nervous system functions. Some polypores have been used in ritual and for utilitarian purposes for ages; the famous Ötzi the Iceman was found carrying two different polypore species: Piptoporus betulinus, notable for its long usage in European folk medicines, and Fomes fomentarius, which was likely used for starting fires.

Medicinal mushroom polypores in use today are Ganoderma lucidum coll. (reishi or lingzhi), Trametes versicolor (turkey tail) and Ganoderma applanatum (Japanese Kufuki-saru-no-koshikake). Beyond their traditional use in herbal medicine, contemporary research has suggested many applications of polypores for the treatment of illnesses related to the immune system and cancer recovery.

Several species have been studied for their ability to produce compounds with anti-pathogenic activity.

MORCHELLA

Morchella, the true morels, is a genus of edible sac fungi closely related to anatomically simpler cup fungi in the order Pezizales (division Ascomycota). These distinctive fungi have a honeycomb appearance due to the network of ridges with pits composing their caps. Morels are prized by gourmet cooks, particularly in French cuisine.

Typified by Morchella esculenta in 1794, the genus has been the source of considerable taxonomical controversy throughout the years, mostly with regard to the number of species involved, with some mycologists recognising as few as three species and others over thirty. Current molecular phylogenetics suggest there might be over seventy species of Morchella worldwide, most of them exhibiting high continental endemism and provincialism.

The genus is currently the focus of extensive phylogenetic, biogeographical, taxonomical and nomenclatural studies, and several new species have been described from Australia, Canada, Cyprus, Israel, North America, Spain, and Turkey.

A black morel in Poland.

Ecology, Phenology and Distribution

The ecology of Morchella species is not well understood. Many species appear to form symbiotic or endophytic relationships with trees, while others appear to act as saprotrophs. Yellow morels (Morchella esculenta and related species) are more commonly found under deciduous trees rather than conifers, while black morels (Morchella elata and related species) are mostly found in coniferous forests, disturbed ground and recently burned areas. Morchella galilaea, and occasionally Morchella rufobrunnea, appear to fruit in the autumn or winter months rather than spring, which is the typical fruiting season for morels.

Yellow morels in West Virginia, USA.

Tree species associated with Morchella vary greatly depending on the individual species, continent, or region. Trees commonly associated with morels in Europe and across the Mediterranean include Abies (fir), Pinus (pine), Populus (poplar), Ulmus (elm), Quercus (oak), Arbutus (strawberry trees), Castanea (chestnut), Alnus (alder), Olea (olive trees), Malus (apple trees) and Fraxinus (ash). In western North America morels are often found in coniferous forests, including species of Pinus (pine), Abies (fir), Larix (larch), and Pseudotsuga (Douglas-fir), as well as in Populus (cottonwood) riparian forests. Deciduous trees commonly associated with morels in the northern hemisphere include Fraxinus (ash), Platanus (sycamore), Liriodendron (tulip tree), dead and dying elms, cottonwoods and old apple trees (remnants of orchards). Due to their springtime phenology (March–May), morels are hardly ever found in the vicinity of common poisonous mushrooms such as the death cap (Amanita phalloides), the sulphur tuft (Hypholoma fasciculare), or the fly agaric (Amanita muscaria). They can, however, occur alongside false morels (Gyromitra and Verpa species) and elfin saddles (Helvella species), which also appear in spring.

Efforts to cultivate morels at a large scale have rarely been successful and the commercial morel industry relies on the harvest of wild mushrooms.

Association with Wildfire

Certain Morchella species (M. eximia, M. importuna, M. tomentosa and others) exhibit a pyrophilic behaviour and may grow abundantly in forests which have been recently burned by a fire. Moderate-intensity fires are reported to produce higher abundances of morels than low- or high-intensity fires. This is caused by the soil becoming more alkaline as the result of wood ash combining with water and being absorbed into the soil which triggers the morels to fruit. Alkaline soil conditions which trigger fruiting has been observed and exploited with small scale commercial cultivation of morels. Where fire suppression is practiced, morels often grow in small amounts in the same spot year after year. If these areas are overrun by wildfire they often produce a bumper crop of black morels the following spring. Commercial pickers and buyers in North America target recently burned areas for this reason. These spots may be closely guarded by mushroom pickers, as morels are widely regarded as a delicacy and often a cash crop.

Longneck morel in Indiana, USA.

Transcontinental Species

Although the majority of species within Morchella exhibit high continental endemism and provincialism, several species have been phylogenetically shown to be present in more than one continent. So far, the list of transcontinental species includes M. americana, M. eohespera, M. eximia, M. exuberans, M. galilaea, M. importuna, M. populiphila, M. pulchella, M. rufobrunnea, M. semilibera, M. sextelata, M. steppicola and M. tridentina. The reasons behind the widespread, cosmopolitan distribution of these species, are still puzzling. Long-distance spore dispersal has also been suggested as a possible dispersal mechanism for some species, especially those belonging to fire-adapted lineages.

Black morel.

In Popular Culture

Morel hunting is a common springtime activity. Mushroom collectors may carry a mesh collecting bag, so the spores can scatter as one carries the harvest.

Every spring, hundreds of morel enthusiasts gather in Boyne City, Michigan for the National Morel Mushroom Festival, a century old event. As one observer stated, "if there is a modern, North American reenactment of Geoffrey Chaucer's Canterbury Tales this is it." Other festivals and hunting competitions in North America include the Illinois State Morel Mushroom Hunting Championship and the Ottawa Midwest Morel Fest.

Vernacular Names

Morchella species have been called by many local names; some of the more colorful include dryland fish, because when sliced lengthwise then breaded and fried, their outline resembles the shape of a fish; hickory chickens, as they are known in many parts of Kentucky; and merkels or miracles, based on folklore, of how a mountain family was saved from starvation by eating morels. In parts of West Virginia, they are known as molly moochers, muggins, or muggles. Due to the partial structural and textural similarity to some species of Porifera (sponges), another common name for any true morel is sponge mushroom. In the Appalachian woodlands, morels have also been called haystacks, or snakeheads. The Finnish vernacular name huhtasieni, refers to huhta, area cleared for agriculture by the slash and burn method.

Cultivation

Due to the mushroom's prized fruit bodies, several attempts have been made to grow the fungus in culture. In 1901, Repin reported successfully obtaining fruit bodies in a cave in which cultures had been established in flower pots nine years previously in 1892.

More recently, small scale commercial growers have had success growing morels by using partially shaded rows of mulched wood. The rows of mulch piles are inoculated with morel mushroom spores in a solution of water and molasses which are poured over the piles of mulch and then they are allowed to grow undisturbed for several weeks. A solution of wood ashes mixed in water and diluted is subsequently poured over the rows of wood mulch which triggers fruiting of the morels. Morels are known to appear after fires and the alkaline conditions produced by wood ash mixed with water initiate fruit body formation for most species of morels.

Nutrition

A 100g serving of morel mushrooms contains 31 calories, 5.1 grams of carbohydrates, 3.1 grams of protein, 2.8 grams of fiber, and negligible fat. Morel mushrooms are a rich source of iron and vitamin D, with moderate levels of vitamins B_2, B_3, B_6, and manganese, phosphorus and zinc. They have low levels of calcium, potassium, and vitamins B_1, B_5, and folate.

Gastronomical Value and Culinary Uses

Morels are a feature of many cuisines, including Provençal. Their unique flavor is prized by chefs worldwide, with recipes and preparation methods designed to highlight and preserve it. As with most edible fungi, they are best when collected or bought fresh. One of the best and simplest ways to enjoy morels is by gently sauteeing them in butter, cracking pepper on top and sprinkling with salt. They make great additions to meat and poultry dishes, soups, or can be used as pasta fillings. However, as morels are known to contain thermolabile toxins, they must always be cooked before eating.

Yellow morels in France.

Morels can be preserved in several ways: They can be 'flash frozen' by simply running under cold water or putting them in a bucket to soak for a few minutes, then spread on a baking tray and placed into a freezer. After freezing, they keep very well with the frozen glaze for a long time in airtight plastic containers. However, when thawed they can sometimes turn slightly mushy, so they are best frozen after steaming or frying. Due to their natural porosity, morels may contain trace amounts of soil which cannot be easily washed out. Any visible soil should be removed with a brush, after cutting the body in half lengthwise, if needed. Mushroom hunters sometimes recommend soaking morels in a bowl of salt water briefly prior to cooking, although many chefs would disagree.

Drying is a popular and effective method for long-term storage, and morels are widely available commercially in this form. Any insect larvae which might be present in the fruit bodies usually drops out during the drying process. Dried morels can then be reconstituted by soaking for 10–20 minutes in warm water or milk, and the soaking liquid can be used as stock.

The supreme flavour of morels is not just appreciated by humans; in Yellowstone National Park, black morels are also known to be consumed by grizzly bears (Ursus arctos horribilis).

Black morels in British Columbia, Canada.

Toxicity

Morchella species are thought to contain small amounts of hydrazine toxins or an unknown toxin that is destroyed through cooking, (the presence of hydrazine is controversial since there are no primary references of hydrazine having been detected in the species); because of this, morels should never be eaten raw. It has been reported that even cooked morels can sometimes cause symptoms of upset stomach when consumed with alcohol.

When eating this fungus for the first time it is wise to consume a small amount to minimize any allergic reaction. As with all fungi, morels for consumption must be clean and free of decay. Morels growing in old apple orchards that had been treated with the insecticide lead arsenate may accumulate levels of toxic lead and arsenic that are unsuitable for human consumption.

False Morels

Gyromitra esculenta, a false morel.

When gathering morels for the table, care must be taken to distinguish them from the poisonous "false morels", a term loosely applied to describe Gyromitra esculenta, Verpa bohemica, and other morel lookalikes. Although false morels are sometimes eaten without ill effect, they can cause severe gastrointestinal upset, loss of muscular coordination (including cardiac muscle), or even death. Incidents of poisoning usually occur when they are eaten in large quantities, inadequately cooked, or over several days in a row. False morels contain gyromitrin, an organic carcinogenic poison, hydrolyzed in the body into monomethylhydrazine (MMH). Gyromitra esculenta in particular, has been reported to be responsible for up to 23% of mushroom fatalities each year in Poland.

The key morphological features distinguishing false morels from true morels are as follows:

- Gyromitra species often have a "wrinkled" or "cerebral" (brain-like) appearance to the cap due to multiple wrinkles and folds, rather than the honeycomb appearance of true morels due to ridges and pits.

- Gyromitra esculenta has a cap that is usually reddish-brown in colour, but sometimes also chestnut, purplish-brown, or dark brown.
- Gyromitra species are typically chambered in longitudinal section, while Verpa species contain a cottony substance inside their stem, in contrast to true morels which are always hollow.
- The caps of Verpa species (V. bohemica, V. conica and others) are attached to the stem only at the apex (top of the cap), unlike true morels which have caps that are attached to the stem at, or near the base of the cap. The easiest way to distinguish Verpa species from Morchella species is to slice them longitudinally.

TRUFFLE

A truffle is the fruiting body of a subterranean ascomycete fungus, predominantly one of the many species of the genus Tuber. In addition to Tuber, many other genera of fungi are classified as truffles including Geopora, Peziza, Choiromyces, Leucangium, and over a hundred others. These genera belong to the class Pezizomycetes and the Pezizales order. There are several truffle-like basidiomycetes excluded from Pezizales including Rhizopogon and Glomus. Truffles are ectomycorrhizal fungi and are therefore usually found in close association with tree roots. Spore dispersal is accomplished through fungivores, animals that eat fungi. These fungi have significant ecological roles in nutrient cycling and drought tolerance.

Black truffle (Tuber melanosporum).

Some of the truffle species are highly prized as food. French gourmet Jean Anthelme Brillat-Savarin called truffles "the diamond of the kitchen". Edible truffles are held in high esteem in French, Italian, Croatian, Slovene, Ottoman, Middle Eastern and Spanish cuisine, as well as in international haute cuisine. Truffles are cultivated agriculturally and are also harvested from natural habitats.

White truffles from San Miniato.

Black truffles, San Miniato.

Phylogeny and Species

Phylogenetic analysis has demonstrated the convergent evolution of the ectomycorrhizal trophic mode in diverse fungi. The subphylum, Pezizomycotina, containing the order Pezizales, is approximately 400 million years old. Within the order Pezizales, subterranean fungi evolved independently at least fifteen times. Contained within Pezizales are the families Tuberaceae, Pezizaceae, Pyronematacae, and Morchellaceae. All of these families contain lineages of subterranean or truffle fungi. The oldest ectomycorrhizal fossil is from the Eocene about 50 million years ago. This indicates that the soft bodies of ectomycorrhizal fungi do not easily fossilize. Molecular clockwork has suggested the evolution of ectomycorrhizal fungi occurred approximately 130 million years ago.

The evolution of subterranean fruiting bodies has arisen numerous times within the Ascomycota, Basidiomycota, and Glomeromycota. For example, the genera Rhizopogon and Hysterangium of Basidiomycota both form subterranean fruiting bodies and play similar ecological roles as truffle forming ascomycetes. The ancestors of the Ascomycota genera Geopora, Tuber, and Leucangium originated in Laurasia during the Paleozoic era. Phylogenetic evidence suggests that the majority of subterranean fruiting bodies evolved from above-ground mushrooms. Over time mushroom stipes and caps were reduced, and caps began to enclose reproductive tissue. The dispersal of sexual spores then shifted from wind and rain to utilizing animals.

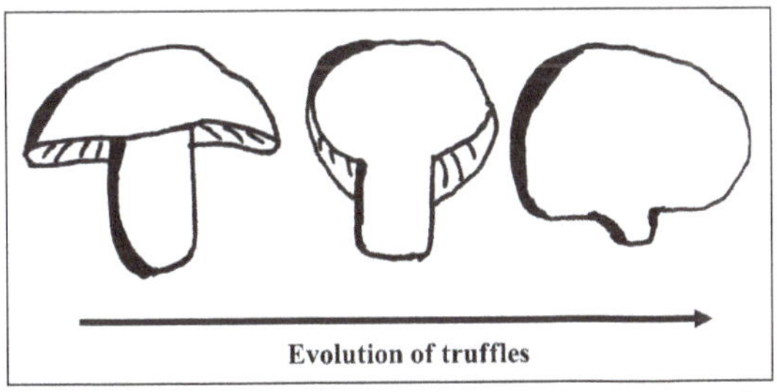

Evolution of subterranean fruiting bodies from above-ground mushrooms.

The phylogeny and biogeography of the genus Tuber was investigated in 2008 using internal transcribed spacers (ITS) of nuclear DNA and revealed five major clades (Aestivum, Excavatum, Rufum, Melanosporum and Puberulum); this was later improved and expanded in 2010 to nine major clades using large subunits (LSU) of mitochondrial DNA. The Magnatum and Macrosporum clades were distinguished as distinct from the Aestivum clade. The Gibbosum clade was resolved as distinct from all other clades, and the Spinoreticulatum clade was separated from the Rufum clade.

The truffle habit has evolved independently among several basidiomycete genera. Phylogenetic analysis has revealed that basidiomycete subterranean fruiting bodies, like their ascomycete counterparts, evolved from above ground mushrooms. For example, it is likely that Rhizopogon species arose from an ancestor shared with Suillus, a mushroom forming genus. Studies have suggested that selection for subterranean fruiting bodies among ascomycetes and basidiomycetes occurred in water-limited environments.

Black Truffle

Black Périgord truffle, cross-section.

The black truffle or black Périgord truffle (Tuber melanosporum), the second-most commercially valuable species, is named after the Périgord region in France. Black truffles associate with oaks, hazelnut, cherry, and other deciduous trees and are harvested in late autumn and winter. The genome sequence of the black truffle was published in March 2010.

Summer or Burgundy Truffle

The black summer truffle (Tuber aestivum) is found across Europe and is prized for its culinary value. Burgundy truffles (designated Tuber uncinatum, but the same species) are harvested in autumn until December and have aromatic flesh of a darker colour. These associate with various trees and shrubs.

Summer truffles in a shop in Rome.

White Truffle

White truffle washed and corner cut to show interior.

Tuber magnatum, the high-value white truffle or trifola d'Alba Madonna ("Truffle of the White Madonna" in Italian) is found mainly in the Langhe and Montferrat areas of Italy of the Piedmont region in northern Italy and, most famously, in the countryside around the cities of Alba and Asti. A large percentage of Italy's white truffles also come from Molise.

Whitish Truffle

The "whitish truffle" (*Tuber borchii*) is a similar species found in Tuscany, Abruzzo, Romagna, Umbria, the Marche and Molise. It is not as aromatic as those from Piedmont, although those from Città di Castello come quite close.

Geopora Species

Geopora spp. are important ectomycorrhizal partners of trees in woodlands and forests throughout the world. Pinus edulis, a widespread pine species of the Southwest, is dependent on Geopora for nutrient and water acquisition in arid environments. Like other truffle fungi, Geopora produces subterranean sporocarps as a means of sexual reproduction. Geopora cooperi, also known as pine truffle or fuzzy truffle, is an edible species of this genus.

Other Species

A less common truffle is "garlic truffle" (Tuber macrosporum).

In the U.S. Pacific Northwest, several species of truffle are harvested both recreationally and commercially, most notably, the Leucangium carthusianum, Oregon black truffle; Tuber gibbosum, Oregon spring white truffle; and Tuber oregonense, the Oregon winter white truffle. Kalapuya brunnea, the Oregon brown truffle, has also been commercially harvested and is of culinary note.

The pecan truffle (Tuber lyonii) syn. texense is found in the Southern United States, usually associated with pecan trees. Chefs who have experimented with them agree "they are very good and have potential as a food commodity". Although pecan farmers used to find them along with pecans and discard them, considering them a nuisance, they sell for about $160 a pound and have been used in some gourmet restaurants.

Truffle-like Species

The term "truffle" has been applied to several other genera of similar underground fungi. The genera Terfezia and Tirmania of the family Terfeziaceae are known as the "desert truffles" of Africa and the Middle East. The basidiomycete "Hart's truffle" is a name for Elaphomycetaceae. Pisolithus tinctorius, which was historically eaten in parts of Germany, is sometimes called "Bohemian truffle".

Rhizopogon spp. are ectomycorrhizal members of the Basidiomycota and the order Boletales, a group of fungi that typically form mushrooms. Like their ascomycete counterparts, these fungi are capable of creating truffle-like fruiting bodies. Rhizopogon spp. are ecologically important in coniferous forests where they associate with various pines, firs, and Douglas-fir. In addition to their ecological importance, these fungi hold economic value as well. Rhizopogon spp. are commonly used to inoculate coniferous seedlings in nurseries and during reforestation.

Rhizopogon truffle.

Hysterangium spp. are ectomycorrhizal members of the Basidiomycota and the order Hysterangiales that form sporocarps similar to true truffles. These fungi form mycelial mats of vegetative hyphae that may cover 25-40% of the forest floor in Douglas-fir forests, thereby contributing to a significant portion of the biomass present in soils. Like other ectomycorrhizal fungi, Hysterangium play a role in nutrient exchange in the nitrogen cycle by accessing nitrogen unavailable to host plants and by acting as nitrogen sinks in forests.

Glomus spp. are arbuscular mycorrhizae of the phylum Glomeromycota within the order Glomerales. Members of this genus have low host specificity, associating with a variety of plants including hardwoods, forbs, shrubs and grasses. These fungi commonly occur throughout the Northern hemisphere.

Ecology

The mycelia of truffles form symbiotic, mycorrhizal relationships with the roots of several tree species including beech, birch, hazel, hornbeam, oak, pine, and poplar. Mutualistic ectomycorrhizal fungi such as truffles provide valuable nutrients to plants in exchange for carbohydrates. Ectomycorrhizal fungi lack the ability to survive in the soil without their plant host. In fact, many of these fungi have lost the enzymes necessary for obtaining carbon through other means. For example, truffle fungi have lost their ability to degrade the cell walls of plants limiting their capacity to decompose plant litter. Plant hosts can also be dependent on their associated truffle fungi. Studies have demonstrated that Geopora, Peziza, and Tuber spp. are vital in the establishment of oak communities.

Tuber species prefer argillaceous or calcareous soils that are well drained and neutral or alkaline. Tuber truffles fruit throughout the year, depending on the species, and can be found buried between the leaf litter and the soil. The majority of fungal biomass is found in the humus and litter layers of soil.

Most truffle fungi produce both asexual spores (mitospores or conidia) and sexual spores (meiospores or ascospores/basidiospores). Conidia can be produced more readily and with less energy than ascospores and can disperse during disturbance events. Production of ascospores is energy intensive because the fungus must allocate resources to the production of large sporocarps. Ascospores are borne within sac-like structures called asci, which are contained within the sporocarp.

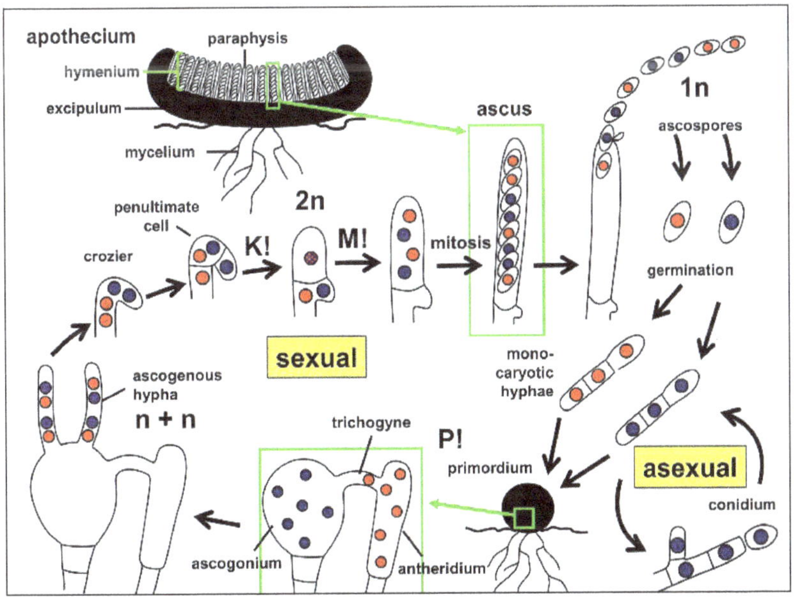

Life cycle of the order Pezizales in Ascomycota.

Because truffle fungi produce their sexual fruiting bodies underground, spores cannot be spread via wind and water. Therefore, nearly all truffles depend on mycophagous animal vectors for spore dispersal. This is analogous to the dispersal of seeds in fruit of angiosperms. When the ascospores are fully developed, the truffle will begin to exude volatile compounds that serve to attract animal vectors. For successful dispersal, these spores must survive passage through the digestive tracts of animals. Ascospores have thick walls composed of chitin to help them endure the environment of animal guts. Animal vectors include birds, deer, and rodents such as voles, squirrels, and chipmunks. Many species of trees, such as Quercus garryana, are dependent on the dispersal of sporocarps to inoculate isolated individuals. For example, the acorns of Q. garryana may be carried to new territory that lacks the necessary mycorrhizal fungi for establishment. Some mycophagous animals depend on truffles as their dominant food source. Flying squirrels, Glaucomys sabrinus, of North America play a role in a three-way symbiosis with truffles and their associated plants. Glaucomys sabrinus is particularly adapted to finding truffles using its refined sense of smell, visual clues, and long-term memory of prosperous populations of truffles. This intimacy between animals and truffles indirectly influences the success of mycorrhizal plant species.

After ascospores are dispersed, they remain dormant until germination is initiated by exudates excreted from host plant roots. Following germination, hyphae will form and seek out the roots of host plants. Arriving at roots, hyphae will begin to form a mantle or sheath on the outer surface of root tips. Hyphae will then enter the root cortex intercellularly to form the Hartig net for nutrient exchange. Hyphae can spread to other root tips colonizing the entire root system of the host. Over time, the truffle fungus accumulates sufficient resources to form fruiting bodies. Rate of growth is correlated with increasing photosynthetic rates in the spring as trees leaf out.

Nutrient Exchange

In exchange for carbohydrates, truffle fungi provide their host plants with valuable micro and macronutrients. Plant macronutrients include potassium, phosphorus, nitrogen, and sulfur whereas micronutrients include iron, copper, zinc, and chloride. In truffle fungi, as in all ectomycorrhizae, the majority of nutrient exchange occurs in the Hartig net, the intercellular hyphal network between plant root cells. A unique feature of ectomycorrhizal fungi is the formation of the mantle on outer surface of fine roots.

Truffles have been suggested to co-locate with the orchid species Epipactis helleborine and Cephalanthera damasonium., though this is not always the case.

Nutrient Cycling

Truffle fungi are ecologically important in nutrient cycling. Plants obtain nutrients via their fine roots. Mycorrhizal fungi are much smaller than fine roots and therefore have a higher surface area and a greater ability to explore soils for nutrients. Acquisition of nutrients includes the uptake of phosphorus, nitrogen, iron, magnesium and other ions. Many ectomycorrhizal fungi form fungal mats in the upper layers of soils surrounding host plants. These mats have significantly higher concentrations of carbon and fixed nitrogen than surrounding soils. Because these mats are nitrogen sinks, leaching of nutrients is reduced. Mycelial mats can also help maintain the structure of soils by holding organic matter in place and preventing erosion. Often these networks of mycelium provide support for smaller organisms in the soil, such as bacteria and microscopic arthropods. Bacteria feed on the exudates released by mycelium and colonize soil surrounding them. Microscopic arthropods such as mites feed directly on mycelium and release valuable nutrients for the uptake of other organisms. Thus, truffle fungi, along with other ectomycorrhizal fungi, facilitate a complex system of nutrient exchange between plants, animals, and microbes.

Importance in Arid-land Ecosystems

Plant community structure is often affected by the availability of compatible mycorrhizal fungi. In arid-land ecosystems, these fungi become essential for the survival of their host plants by enhancing ability to withstand drought. A foundation species in arid-land ecosystems of the Southwest United States is Pinus edulis, commonly known as pinyon pine. Pinus edulis associates with the subterranean fungi Geopora and Rhizopogon. As global temperatures rise, so does the occurrence of severe droughts detrimentally affecting the survival of arid-land plants. This variability in climate has increased the mortality of P. edulis. Therefore, the availability of compatible mycorrhizal inoculum can greatly affect the successful establishment of P. edulis seedlings. Associated ectomycorrhizal fungi will likely play a significant role in the survival of P. edulis with continuing global climate change.

Extraction

Table: Comparison of truffle dog and hog.

Truffle Dog	Truffle Hog
Keen sense of smell.	Keen sense of smell.
Must be trained.	Innate ability to sniff out truffles.
Easier to control.	Tendency to eat truffles once found.

Trained pig in Gignac, Lot, France.

Because truffles are subterranean, they are often located with the help of an animal possessing a refined sense of smell. Traditionally, pigs have been utilized for the extraction of truffles. Both the female pig's natural truffle-seeking, as well as her usual intent to eat the truffle, are due to a compound within the truffle similar to androstenol, the sex pheromone of boar saliva, to which the sow is keenly attracted. Studies in 1990 demonstrated that the compound actively recognized by both truffle pigs and dogs is dimethyl sulfide.

In Italy, the use of the pig to hunt truffles has been prohibited since 1985 because of damage caused by animals to truffle mycelia during the digging that dropped the production rate of the area for some years. An alternative to truffle pigs are dogs. Dogs pose an advantage in that they do not have a strong desire to eat truffles and can therefore be trained to locate sporocarps without digging them up. Pigs will attempt to dig up truffles.

Trained dog in Mons, Var, France.

Fly species of the genus Suilla can also detect the volatile compounds associated with subterranean fruiting bodies. These flies will lay their eggs above truffles to provide food for their young. At ground level Suilla can be seen flying above truffles.

Volatile Constituents

The volatile constituents responsible for the natural aroma of truffles are released by the mycelia, fruiting body or derive from truffle-associated microbes. The chemical ecology of truffle volatiles is complex, interacting with plants, insects and mammals, which contribute to spore dispersal. Depending on the truffle species, life cycle or location, they include:

- Sulfur volatiles, which occur in all truffle species, such as dimethyl mono- (DMS), di- (DMDS) and tri- (DMTS) sulfides, as well as 2-methyl-4,5-dihydrothiophene, characteristic of the white truffle T. borchii and 2,4-Dithiapentane occurring in all species but mostly characteristic of the white truffle T. magnatum. Some of the very aromatic white truffles are notably pungent, even irritating the eye when cut or sliced.
- Metabolites of non-sulfur amino acid constituents (simple and branched chain hydrocarbons) such as ethylene (produced by mycelia of white truffles affecting

root architecture of host tree), as well as 2-methylbutanal, 2-methylpropanal and 2-phenylethanol (also common in baker's yeast).

- Fatty acid-derived volatiles (C8-alcohols and aldehydes with a characteristic fungal odor, such as 1-octen-3-ol and 2-octenal). The former is derived from linoleic acid, and produced by mature white truffle T. borchii.
- Thiophene derivatives appear to be produced by bacterial symbionts living in the truffle body. The most abundant of these, 3-methyl,4-5 dihydrothiophene, contributes to white truffle's aroma.

A number of truffle species and varieties are differentiated based on their relative contents or absence of sulfides, ethers or alcohols, respectively. The sweaty-musky aroma of truffles is similar to that of the pheromone androstenol that also occurs in humans. As of 2010, the volatile profiles of seven black and six white truffle species have been studied.

Culinary Use

Because of their high price and their pungent aroma, truffles are used sparingly. Supplies can be found commercially as unadulterated fresh produce or preserved, typically in a light brine.

As the volatile aromas dissipate quicker when heated, truffles are generally served raw and shaved over warm, simple foods where their flavor will be highlighted, such as buttered pasta or eggs. Thin truffle slices may be inserted into meats, under the skins of roasted fowl, in foie gras preparations, in pâtés, or in stuffings. Some specialty cheeses contain truffles as well. Truffles are also used for producing truffle salt and truffle honey.

While in the past chefs used to peel truffles, in modern times, most restaurants brush the truffle carefully and shave it or dice it with the skin on so as to make the most of the valuable ingredient. Some restaurants stamp out circular discs of truffle flesh and use the skins for sauces.

Truffle oil (olive oil with Tuber melanosporum).

Truffle Oil

Truffle oil is used as a lower-cost and convenient substitute for truffles, to provide flavouring, or to enhance the flavour and aroma of truffles in cooking. Many "truffle oils", however, do not contain any truffles, while others will include pieces of one of the many inexpensive, unprized truffle varietals, which have no culinary value simply for show. The vast majority is oil which has been artificially flavoured using a synthetic agent such as 2,4-dithiapentane.

Truffle Vodka

Because a greater variety of aromatic molecules in truffles are soluble in alcohol, it can be used to carry a more complex and accurate truffle flavour than oil without the need for synthetic flavourings. However, many commercial producers of "truffled" alcohol use 2,4-dithiapentane regardless, as it has become the dominant flavor most consumers, who have never been exposed to fresh truffles but only truffle oils, associate with them. Because most Western nations do not have ingredient labeling requirements for spirits, there is often no way for consumers to know if additional flavorings have been used. Although used as a spirit in its own right and mixed in a range of cocktails, truffle-flavored alcohol is also used by chefs to flavour dishes.

YEAST

Yeasts are a phylogenetically diverse grouping of single-celled fungi. As members of the Kingdom Fungi, which also includes mushrooms, molds, and mildews, yeasts are eukaryotes (organisms with a distinct, membrane-bound nucleus) that digest their food externally and absorb the nutrient molecules into their cells. Yeasts do not form a specific taxonomic or phylogenetic grouping; they are a diverse assemblage of unicellular organisms that occur in two divisions of fungi (Ascomycotina and Basidiomycotina) and 100 genera. Most reproduce asexually by budding, although in a few cases by binary fission. Yeasts that reproduce via budding are known as "true yeasts" and are classified in the order Saccharomycetales. Although yeasts are unicellular, some species with yeast forms may become multicellular through the formation of a string of connected budding cells known as pseudohyphae, or true hyphae as seen in most molds.

The term yeast is also used for commercial baking products that consist of yeast or dried yeast cells. The term yeast is often taken as a synonym for S. cerevisiae, or "baker's yeast".

Yeast species demonstrate an incredible diversity of forms. Although approximately 1,500 species have been described, it is estimated that this number represents only 1 percent of all yeast species. The size of these microorganisms varies greatly depending on the species; yeast cells typically measure 3 to 7 micrometers in diameter, although some can reach over 40 μm.

Since they derive energy from fermentation (a chemical conversion of carbohydrates to ethanol and carbon dioxide), yeasts are found primarily in sugar-rich environments, such as fruit surfaces or flower nectars. In advancing their own need for survival, maintenance, and reproduction, they also provide value for ecosystems, as decomposers, and to humans. Their ability to ferment sugars has made them an integral component of human civilization: the yeast species Saccharomyces cerevisiae, in particular, has been used in the baking of leavened bread and the production of alcoholic beverages for thousands of years. Recently, yeasts have been used to generate electricity in microbial fuel cells and to produce ethanol for the biofuel industry.

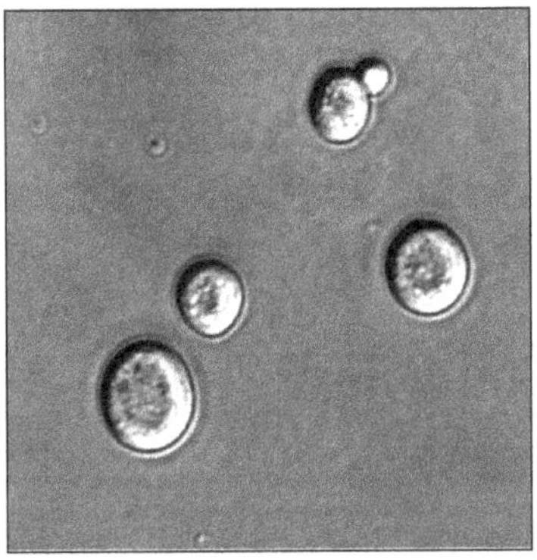

Yeast of the species Saccharomyces cerevisiae.

Yeasts also are important as model organisms in modern cell biology research. S. cerevisiae was the first eukaryote whose genome (the complete DNA sequence of a set of chromosomes) was fully mapped. Given the extensive similarities between yeast cells and human cells, yeast homologues (or counterparts) of human genes, particularly those associated with specific diseases, may provide insights that aid in medical diagnosis and treatment.

Some species of yeast, such as Candida albicans, are opportunistic pathogens that can cause infection in humans, particularly those with compromised immune systems.

Growth and Nutrition

Yeasts are chemoorganotrophs: They use organic compounds as a source of energy and do not require light to grow. They obtain their main source of carbon from hexose (six-carbon) sugars such as glucose, or disaccharides such as sucrose and maltose. Some species can metabolize five-carbon (pentose) sugars such as fructose, as well as alcohols and organic acids. Species that require oxygen for cellular respiration are an example of obligate aerobes; others that are anaerobic but also have aerobic methods of energy production are called facultative anaerobes. Unlike bacteria, there are no known yeast species that grow only in the absence of oxygen (anaerobically).

Yeasts are ubiquitous in the environment, but are most frequently isolated from sugar-rich media, such as fruits and berries (e.g., grapes, apples, or peaches) and exudates from plants (such as plant saps or cacti). Some yeasts are found in soils and in association with insects. Yeasts are generally grown in the laboratory on solid growth media or liquid broths.

Reproduction

The most common mode of vegetative growth in yeast is asexual reproduction by budding or fission. In binary fission, the parent cell divides into two equal cells. Most yeasts reproduce by budding, during which a small protrusion or bud, the daughter cell, is formed on the parent cell. The nucleus of the parent cell splits into a daughter nucleus and migrates into the daughter cell. The bud continues to grow until it separates from the parent cell, forming a new cell. The bud can develop on different areas of the parent cell depending on the genus of the yeast.

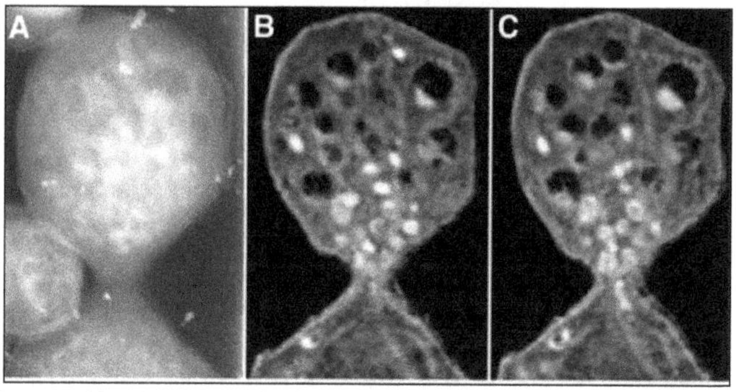

A high magnification view of a budding yeast. Budding, an asexual form of reproduction common among yeast species, entails the formation of a new organism from a protrusion on its parent cell.

Although rare, some species of yeast reproduce via sexual reproductive cycles. Yeast cells can exist as either diploids (which contain two homologous sets of chromosomes) or haploids (cells containing only one set of chromosomes). The mating (or conjugation) of yeast occurs only between haploids, which can be either the a or α (alpha) mating type (thus displaying simple sexual differentiation). Under high stress conditions, such as nutrient depletion, haploid yeast cells will generally die. However, under the same conditions, diploid cells can undergo sporulation, producing a variety of haploid spores. These spores can go on to reproduce sexually, reforming the diploid cell.

Uses

Food and Beverage Production

Yeasts feed on sugars, generating the end-products ethanol (essential to beer and wine production) and carbon dioxide (essential to the baking of leavened bread).

Beer

Beer brewers classify yeasts as top-fermenting and bottom-fermenting. Top-fermenting yeasts are so-called because they form a foam at the top of the wort (the liquid produced from crushed malted grain and water). Top-fermenting yeasts can produce high ethanol concentrations and prefer high temperatures, yielding fruitier, sweeter, ale-type beers than bottom-fermenting yeasts, which are used to produce lager-type beers. These yeasts ferment more sugars, leaving a crisper taste, and grow well at low temperatures.

To ensure purity of strain in industrial brewing, a "clean" sample of the yeast is stored and refrigerated in a laboratory. After a certain number of fermentation cycles, a full scale propagation is produced from this laboratory sample. Typically, it is grown in about three or four stages using sterile brewing wort and oxygen.

Fermenting tanks with yeast being used to brew beer.

Root beer and sodas can be produced using similar methods. The carbonation process created by the active yeast is halted at an earlier stage, generating only trace amounts of alcohol and leaving a significant amount of sugar in the drink.

Wine

Grapes covered in yeast growth observable as a white film, also known as the blush.

In winemaking, yeast converts the sugars present in grape juice or must (juice of freshly pressed grapes that contains various quantities of pulp, skins, stems, and seeds).

Typically, yeast is already present on the grapes, often visible as a powdery film (also known as the bloom or blush) on their exterior. The fermentation can be done with this indigenous (or wild) yeast; however, this may yield unpredictable results depending on the yeast species that are present. For this reason, a pure yeast culture is generally added to the must, which rapidly predominates the fermentation as it proceeds, ensuring a reliable fermentation.

Distilled Beverages

A distilled beverage contains ethanol that has been purified by distillation (method of separating chemical substances based on differences in their volatilities). Carbohydrate-containing plant material is fermented by yeast, producing a dilute solution of ethanol in the process. Spirits such as whiskey and rum are prepared by distilling these dilute solutions of ethanol. Components other than ethanol are collected in the condensate, including water, esters, and other alcohols that account for the flavour of the beverage.

Baking

Yeast is used in baking as a leavening agent, converting the fermentable sugars present in the dough into carbon dioxide. This process causes the dough to rise as the carbon dioxide forms pockets (or bubbles). When the dough is baked, it "sets" and the pockets remain, giving the baked product a soft and spongy texture. The use of potatoes, water from potato boiling, eggs, or sugar in a bread dough accelerates the growth of yeasts. Salt and fats such as butter slow down yeast growth. The majority of yeast used in baking is of the same species (S. cerevisiae) common in alcohol fermentation.

Bread showing pockets left by carbon dioxide.

Kombucha

Yeast Extract

Yeast extract is the common name for various forms of processed yeast products that are used as food additives or flavorings. The general method for making yeast extract for food products such as Vegemite and Marmite on a commercial scale is to add salt to a suspension of yeast, making the solution hypertonic, which leads the cells to shrivel up. This contraction triggers autolysis, during which the yeast's digestive enzymes break down their own proteins into simpler compounds, a self-destructive process. The dying yeast cells are then heated to complete their breakdown, after which the husks (yeasts with thick cell walls) are separated. Yeast autolysates are used in Vegemite and Promite (Australia), Marmite and Bovril (the United Kingdom and Republic of Ireland), Oxo (South Africa, United Kingdom, and Republic of Ireland), and Cenovis (Switzerland).

A Kombucha culture fermenting in a jar.

Nutritional Supplements

Yeast is used in nutritional supplements popular with vegans and the health conscious; in this context, it is often referred to as nutritional yeast. It is a deactivated yeast, usually of the species S. cerevisiae. Nutritional yeast is an excellent source of protein and vitamins, especially the B-complex vitamins, whose functions are related to metabolism, as well as other minerals and cofactors required for growth. It is also naturally low in fat and sodium. Nutritional yeast has a nutty, cheesy, creamy flavor that makes it popular as an ingredient in cheese substitutes; it is often used in place of parmesan cheese and as a topping for popcorn.

Some probiotic supplements use the yeast Saccharomyces boulardii to maintain and restore the natural flora in the large and small gastrointestinal tract.

Other Applications

Bioremediation

Some yeasts can find potential application in the field of bioremediation, which can be defined as any process that uses microorganisms, fungi, green plants, or their enzymes to return an environment altered by contaminants to its original condition. The species Yarrowia lipolytica, for example, is known to degrade palm oil mill effluent, TNT (an explosive material), and other hydrocarbons such as alkanes, fatty acids, fats, and oils.

Industrial Ethanol Production

The ability of yeast to convert sugar into ethanol has been harnessed by the biotechnology industry to generate ethanol fuel. The process starts by milling a feedstock, such as sugar cane, sweetcorn, or cheap cereal grains, then adding dilute sulfuric acid, or fungal alpha amylase enzymes, to break down the starches into complex sugars. A gluco amylase is then added to degrade the complex sugars further into simple sugars. Yeasts are then added to convert the simple sugars to ethanol, which is distilled off to obtain ethanol up to 96 percent in concentration.

Saccharomyces yeasts have been genetically engineered to ferment xylose, one of the major fermentable sugars present in cellulosic biomasses, such as agriculture residues, paper wastes, and wood chips.

Such a development means that ethanol can be efficiently produced from more inexpensive feedstocks, making cellulosic ethanol fuel a more competitively priced alternative to gasoline fuels.

Yeast as a Model Organism

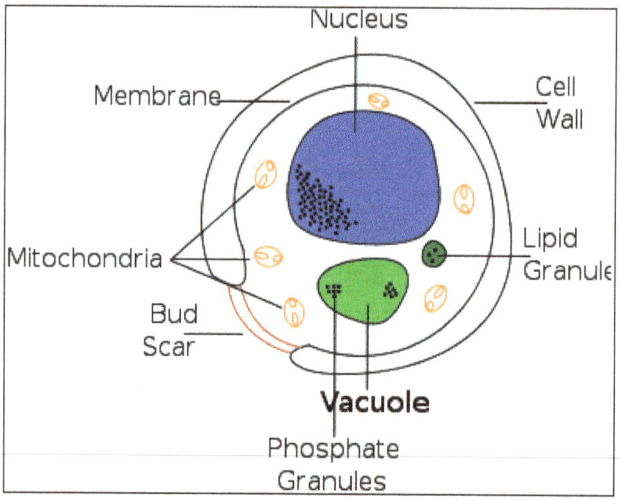

Diagram showing a yeast cell.

Several yeasts, particularly Saccharomyces cerevisiae, have been widely studied as model organisms in genetics and cell biology. The cell cycle in a yeast cell is very similar to the cell cycle in humans; therefore, the basic cellular mechanics of DNA replication, recombination, cell division, and metabolism are comparable. Many proteins important in human biology were first discovered by studying their homologues in yeast; these proteins include cell cycle proteins, signaling proteins, and protein-processing enzymes.

On April 24, 1996, S. cerevisiae was announced to be the first eukaryote to have its genome, consisting of 12 million base pairs, fully sequenced as part of the Genome project. At the time, it was the most complex organism to have its full genome sequenced; the project took seven years and the involvement of over 100 laboratories to accomplish. The second yeast species to have its genome mapped was Schizosaccharomyces pombe; the sequencing was completed in 2002. S. Pombe, whose genome consists of 13.8 million base pairs, was the sixth eukaryotic genome fully sequenced.

Pathogenic Yeasts

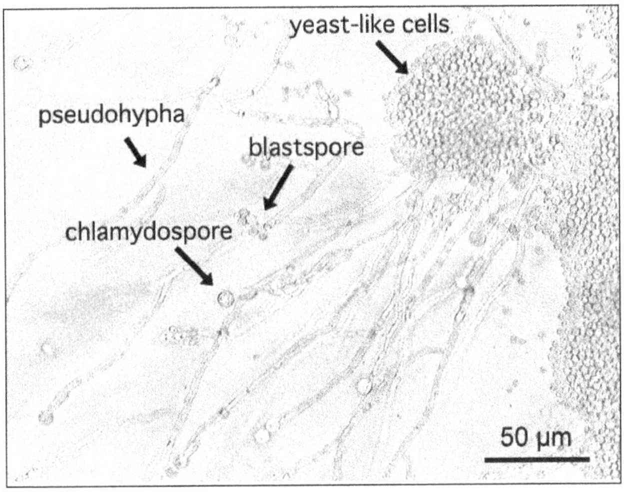

A photomicrograph of Candida albicans showing hyphal outgrowth (a long, branching filamentous cell). Some parasitic fungi like C. albicans use the tip of the hypha to penetrate the host.

Some species of yeast are opportunistic pathogens; that is, they can cause infection in people with compromised immune systems.

Cryptococcus neoformans is a significant pathogen of immunocompromised people, causing the disease cryptococcosis, which occurs in about 7-8% of AIDS patients in the USA, and in a slightly smaller percentage (3-6 percent) in Western Europe. The cells of the yeast are surrounded by a rigid polysaccharide capsule, which helps to prevent them from being recognized and engulfed by white blood cells in the human body. Yeasts of the Candida genus are another group of opportunistic pathogens, causing oral and vaginal infections in humans. Candida is commonly found as a commensal yeast in the mucus membranes of humans and other warm-blooded animals. However, sometimes these strains, which typically do not harm their host, can become pathogenic. In the figure above, the yeast cells sprout a hyphal outgrowth, which locally penetrates the mucosal membrane, causing irritation and shedding of the tissues. Candida glabrata is the second most common Candida pathogen after C. albicans, causing infections of the urogenital tract and of the bloodstream.

References

- Mushroom: newworldencyclopedia.org, Retrieved 08 January, 2020

- Hurtado-Gonzales, O. P.; Valentini, G.; Gilio, T. A.; Martins, A. M.; Song, Q.; Pastor-Corrales, M. A. (2016). "Fine Mapping of Ur-3, a Historically Important Rust Resistance Locus in Common Bean". G3 (Bethesda, Md.). 7 (2): 557–569. doi:10.1534/g3.116.036061. PMC 5295601. PMID 28031244
- Binder, Manfred; Justo, Alfredo; Riley, Robert; Salamov, Asaf; Lopez-Giraldez, Francesc; Sjökvist, Elisabet; Copeland, Alex; Foster, Brian; Sun, Hui; Larsson, Ellen; Larsson, Karl-Henrik; Townsend, Jeffrey; Grigoriev, Igor V.; Hibbett, David S. (2013). "Phylogenetic and phylogenomic overview of the Polyporales". Mycologia. 105 (6): 1350–1373. doi:10.3852/13-003. PMID 23935031
- Loizides, M. (2017). "Morels: the story so far". Field Mycology. 18 (2): 42–53. doi:10.1016/j.fldmyc.2017.04.004
- Læssøe, Thomas; Hansen, Karen (September 2007). "Truffle trouble: what happened to the Tuberales?". Mycological Research. 111 (9): 1075–1099. doi:10.1016/j.mycres.2007.08.004. ISSN 0953-7562. PMID 18022534
- Yeast: newworldencyclopedia.org, Retrieved 13 March, 2020

Reproduction and Spores of Fungi

5 CHAPTER

Fungal reproduction can take place asexually through fragmentation, budding, and producing spores. Some of the fungal spores are sporangiospore, zygospore, ascospore, basidiospore, aeciospore, ascus, oospore, etc. This chapter sheds light on the subject of fungal reproduction and spores for a thorough understanding of it.

FUNGAL REPRODUCTION

Vegetative Reproduction

The most common method of vegetative reproduction is fragmentation. The hypha breaks up into small fragments accidentally or otherwise. Each fragment develops into a new individual. In the laboratory the 'hyphal tip method' is commonly used for inoculation of saprophytic fungus.

In addition to above-mentioned common method of vegetative reproduction the fungi reproduced vegetatively by other means, such as fission, budding, sclerotia, rhizomorphs, etc. In fission, the cell constricts in the centre and divides into two giving rise to new individuals.

The budding is commonly found in Saccharomyces. The buds arise from the protoplasm of the parent cells and ultimately become new individuals.

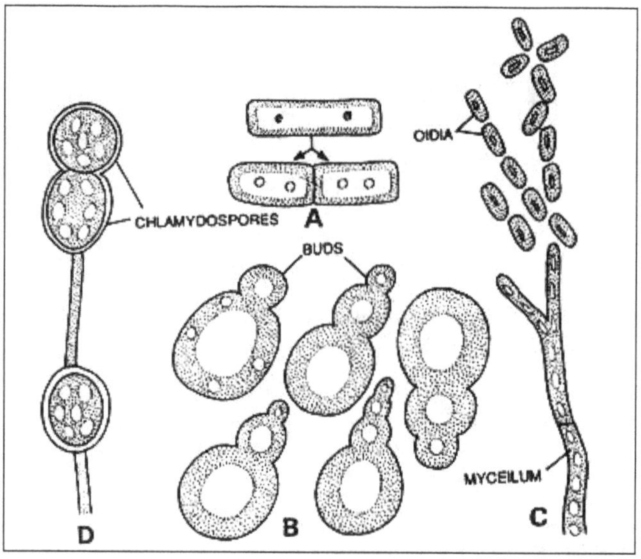

Asexual reproduction: A- Transverse Divicion (fission); B- Budding in yeast Cell; C- Hypha Framenting into oidia or arthrosepores in collybia conige; D- Chlamydospore in fusarium.

The sclerotia are resistant and perennating bodies. They survive for many years. Each sclerotium is cushion-like structure of compact mycelium. They give rise to new mycelia on the approach of favourable conditions.

As mentioned under the modified mycelium, the rope-like rhizomorphs are also resistant to unfavourable conditions and give rise to new mycelia even after several years on the approach of favourable conditions.

Asexual Reproduction

The asexual reproduction takes place by means of spores. Each spore may develop into a new individual. The spores may be produced asexually or sexually and thus named (a) asexual spores and (b) sexual spores. Under asexual reproduction, only asexual spores will be considered.

Asexual Spores

They are innumerable and produced on the diplont mycelium in Phycomycetes and Ascomycetes. In Basidio- mycetes they are produced on the diplont mycelium. The spores are of diverse type and borne upon special structures called the sporophores. These spores are produced asexually and called the asexual spores. Usually the spores are uninucleate and nonmotile but multinucleate and motile spores are also found.

The fungus producing more than one type of spores is called the pleomorphic or polymorphic. The spores produced inside the sporangia are termed the endogenous spores and the spores developing exogenously on the terminal ends of sporophores are called the exogenous spores.

Endogenous Spores

The endogenous spores are produced within the special spore producing cell the sporangium. The sporangia may be terminal or intercalary in their position. The sporophores which bear the sporangia on their apices are called the sporangiophores. They may be branched or unbranched.

The spores produced inside the sporangia are called the endospores or endogenous spores produced inside the sporangia are called the endospores or endogenous spores. They may be motile or non-motile. The motile spores are called the zoospores and the non-motile aplanospores. The zoospores are produced inside the zoosporangia. The protoplasm of the sporangium divides into uninucleate or multinucleate protoplasmic bits and each bit metamorphoses into a spore.

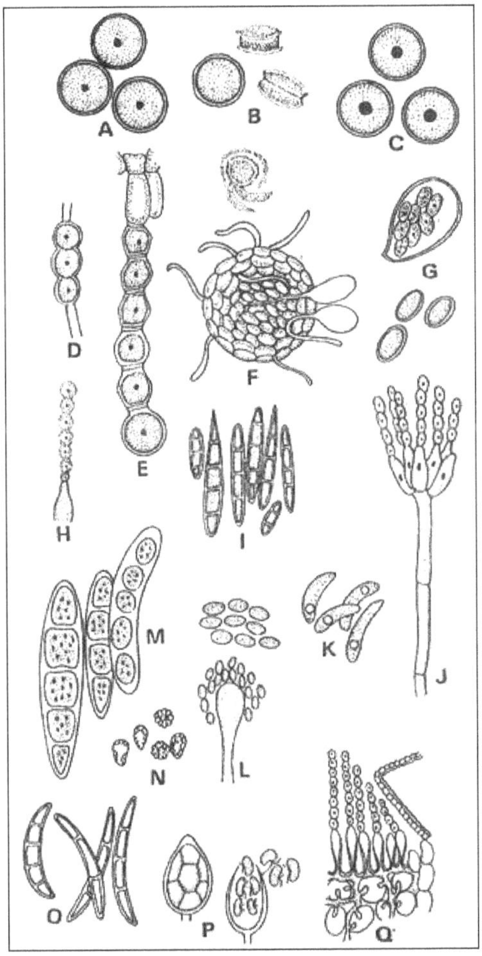

The Fungi, Various types of spores: A- Chlamydospores of Ustilago hordel; B- Ascospores of aspergillus; C- Chlamdospores of ustilag of tritici; D- Chlamydospores of Fusarium; E- Ascospores of erysipher; F- Clestothecium of erysipher; G- Ascus and ascospores of Erysiphe; H- Conidia of Asphergillus; I- Conidia of Cercospora; J- Conidiophore and conidia of penicillium; K- Conidia of colletotrichum sp; L- Aplanos pores of mucor; M- Conidia of Helminthousporium; N- Aplanospores of Rhizopus; O- Conidia of Fusarium sp; P- Sporangium and zoospores fo.

The endogenously produced zoospores are uni or biflagellate. Each spore is without any cell wall, uninucleate and vacuolate. They can move with the help of their flagella. They are usually kidney-shaped or reniform and the flagella are inserted posteriorly or laterally on them. Such zoospores have been recorded from Albugo, Pythium, Phytophthora and many other lower fungi.

The aplanospores are non-motile, without flagella and formed inside the sporangia. They may be uni or multinucleate. These spores lack vacuoles and possess two layered cell walls. The outer thick layer is epispore or exospore which may be ornamented in many cases. The inner thin layer is endospore.

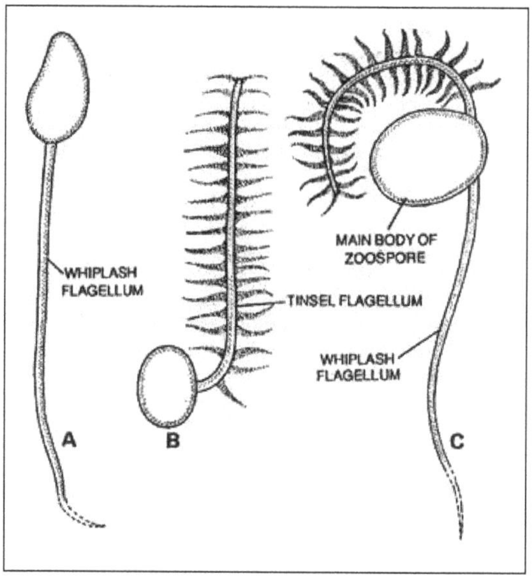

Zoospores: A- Posteriorly unifiagellate with a whiplash flagellum; B- Anteriorly unifagellate with a tinset flagellum; C- Biffagellate with an anterior tinsel and posterior whiplash flagellum.

Exogenous Spores

The spores producing externally or exogenously are either called the exogenous spores or conidia. They are produced externally on the branched or unbranched conidiophores. The condiophores may be septate or aseptate. The conidia borne upon the terminal apices of the conidiophores or the ends of the branches of the condiophores.

The conidia may be produced singly on each sterigma or in chains. The conidial chains may be basipetal to acropetal in succession. The conidia are diverse in their shape and size. They may be unicellular or multicellular, uninucleate or multinucleate. Different genera may be recognized only by the presence of various shaped and various coloured conidia. The conidia of Fungi Imperfecti are multicellular and variously shaped, whereas the conidia of Aspergillus and Penicillium are smoky green coloured and the fungi are called 'the blue-green molds'.

In other type of exospores, the sporophores develop in groups and form the specialized structure called the pustules, pycnia, aecidia, acervuli, and sporodochia. The pycnia are flask-shaped producing pycniospores in them. The acervuli are saucer- shaped widely open bodies having developed conidia in them on small conidiophores.

In mushrooms the sporophores are compactly arranged and form an umbrella-like fructification. The terminal expanded portion bears gills. In each gill there are hundreds of sporophores called the basidia bearing basidiospores. The sporophores (basidia) are arranged in hymenia.

Sexual Reproduction

A large number of fungi reproduce sexually. However, the members of Fungi Imperfecti, or 'Deuteromycetes' lack sexual reproduction.

Usually two phases are found in the life cycle of the plants. These phases are called haploid and diploid phases respectively. The haploid phase possesses the (n) number of chromosomes in the nucleus, whereas this number becomes (2n) in the diploid phase.

The gametes are always haploid (n) and by a sexual fusion they result in diploid (2n) sexual spores, such as zygospores, oospores, etc. To bring haploid (n) phase once again in the life cycle the reduction division (meiosis) takes place and the number of chromosomes becomes half.

The gametes taking part in sexual fusion may be morphologically or physiologically different. Such two gametes taking part in fusion are of opposite sexes or strains, which may be called male and female sex organs or plus and minus strains. When both the sex organs and strains occur on the same mycelicum, the fungus is said to be monoecious or homothallic, and when the male and female sex organs or plus and minus strains occur separately on different mycelia the fungus is said to be dioecious or heterothallic.

The gametes taking part in fusion are usually formed in the cells of sacs called gametangia (singular-gametangium). The morphologically identical male and female gametes are called the isogametes. The morphologically dissimilar male and female gametes are called the heterogametes.

In such cases the male gametes are called the antherozoids and the female ones are the eggs. The fusion of the plasma of the gametes is called the plasmogamy, which is usually followed by the nuclear fusion, i.e., karyogamy. The whole process is called the fertilization.

Sometimes, in some of the fungi, e.g., Phycomycetes and Ascomycetes, the entire contents of the two gametangia fuse with each other, the process is called the gametangial copulation. In the members of Phycomycetes and Ascomycetes the gametangia taking part in gametangial copulation are called the antheridia (singular-antheridium) and the oogonia (singular-oogonium).

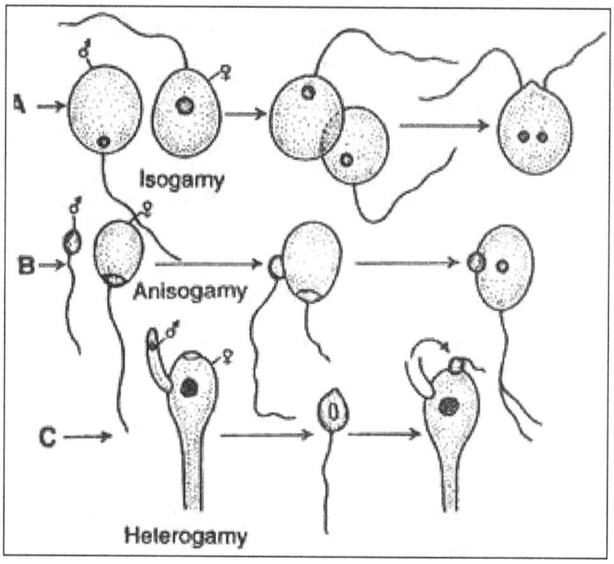

Sexual reproduction in fungi. A, isogamy - as seen in synchytrium, B, anisogamy.

In the lower fungi, there is complete fusion of the nuclei of the two different strained gametes in the sexual union, i.e., karyogamy, whereas in the higher fungi, i.e., Ascomycetes and Basidiomycetes, the fusion of the two nuclei of different strains is delayed and the pairs of the nuclei called the 'dicaryons' are formed. The mycelium having such pairs of nuclei is called the 'dicaryotic mycelium'. In the opposite cases where the mycelium possesses single haploid nucleus of either strain in each cell is called the monocaryotic mycelium. The most common methods of sexual reproduction are as follows:

- Planogametic Copulation: This type of sexual reproduction involves the fusion of two naked gametes one or both of them are motile. The motile gametes are known as planogametes. The most primitive fungi produce insogamous planogametes, e.g., Synchytrium, Plasmodiophorq etc. The anisogamous planogametes are only found in the genus Allomyces of order Blastocladiales. In Monoblepharis (order Monoblepharidales) the unique condition is present here the female gamete is non-motile whereas the male gamete is motile. The male gamete enters the oogonium and fertilizes the egg.

- Gametangial Contact: This method of reproduction is found in many lower fungi (class Phycomycetes). In this method two gametangia of opposite sex (oogonium and antheridium) come in contact and one or more gamete nuclei migrate from the male gametangium (antheridium) to the female gametangium (oogonium).

In no case the gametangia actually fuse. The male nuclei in some species enter the female gametangium through a pore developed by the dissolution of the wall of contact (e.g., in Aspergillus, Penicillium, etc.); in other species the male nuclei migrate through a fertilization tube (e.g., Phythium, Albugo, Peronospora, etc.). After the migration of the nuclei the antheridium eventually disintegrates but the oogonium continues its development in various ways.

- Gametangial Copulation: In this method of sexual reproduction the fusion of the entire contents of two contacting compatible gametangia takes place (e.g., Mucor, Rhizopus, Entomophthora, etc.).

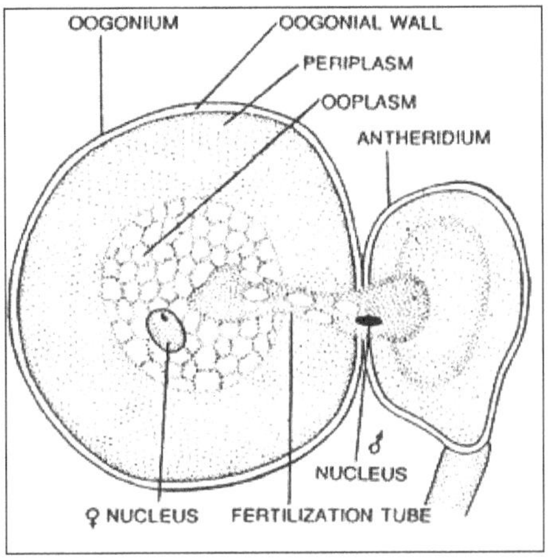

Sexual reproduction: Gametangial contact by means of fertilization tube in pythium aphanidermatuam.

- Spermatization: The sexual reproduction in Neurospora (Class- Ascomycetes) and other fungi takes place by means of this method. The minute, uninucleate, spore-like male structures are known as spermatia. They are produced in several ways. The spermatia are carried by outer agencies to the receptive hyphae (trichogynes) of female gametangia, to which they become attached. A pore develops at the wall of contact and the contents of spermatium pass into the female gametangium through the receptive hypha.

- Somatogamy: The sex organs are not produced. The somatic cells take part in sexual fusion, e.g., Morchella, many higher fungi.

MATING IN FUNGI

Mating in fungi is a complex process governed by mating types. Research on fungal mating has focused on several model species with different behaviour. Not all fungi reproduce sexually and many that do are isogamous; thus, the terms "male" and "female" do not apply to many members of the fungal kingdom. Homothallic species are able to mate with themselves, while in heterothallic species only isolates of opposite mating types can mate.

Mating between isogamous fungi may consist only of a transfer of nuclei from one cell to another. Vegetative incompatibility within species often prevents a fungal isolate from mating with another isolate. Isolates of the same incompatibility group do not mate or mating does not lead to successful offspring. High variation has been reported including same chemotype mating, sporophyte to gametophyte mating and biparental transfer of mitochondria.

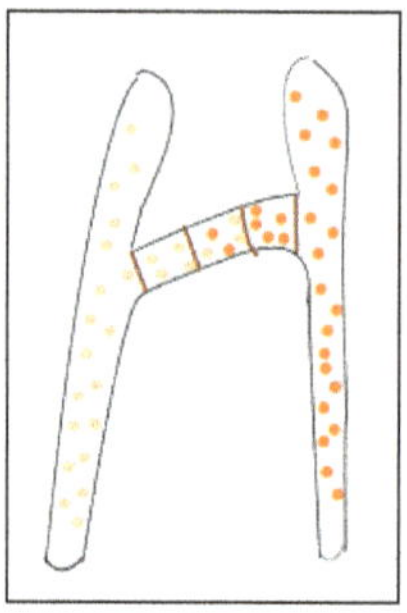

Fungi within Zygomycota form progametangia with suspensors during mating.

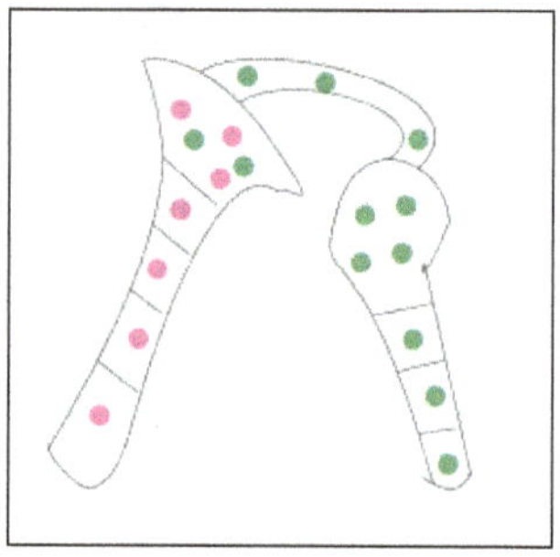

Fungi within Ascomycota form ascogonium and antheridium with trichogyne bridge.

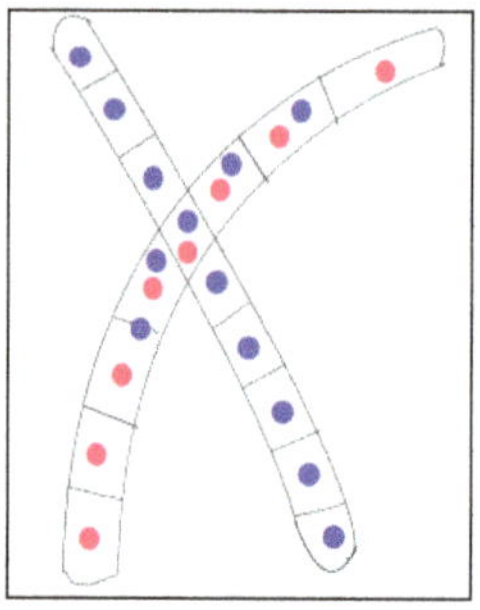

Typical mating fusion of two compatible monokaryons in Basidiomycota.

Mating in Zygomycota

A zygomycete hypha grows towards a compatible mate and they both form a bridge, called a progametangia, by joining at the hyphal tips via plasmogamy. A pair of septa forms around the merged tips, enclosing nuclei from both isolates. A second pair of septa forms two adjacent cells, one on each side. These adjacent cells, called suspensors provide structural support. The central cell is destined to become a spore. The nuclei join in a process called karyogamy to form a zygote.

Mating in Ascomycota

As it approaches a mate, a haploid sac fungus develops one of two complementary organs, a "female" ascogonium or a "male" antheridium. These organs resemble gametangia except that they contain only nuclei. A bridge, the trichogyne forms, that provides a passage for nuclei to travel from the antheridium to the ascogonium. A dikaryote grows from the ascogonium, and karyogamy occurs in the fruiting body.

Neurospora Crassa

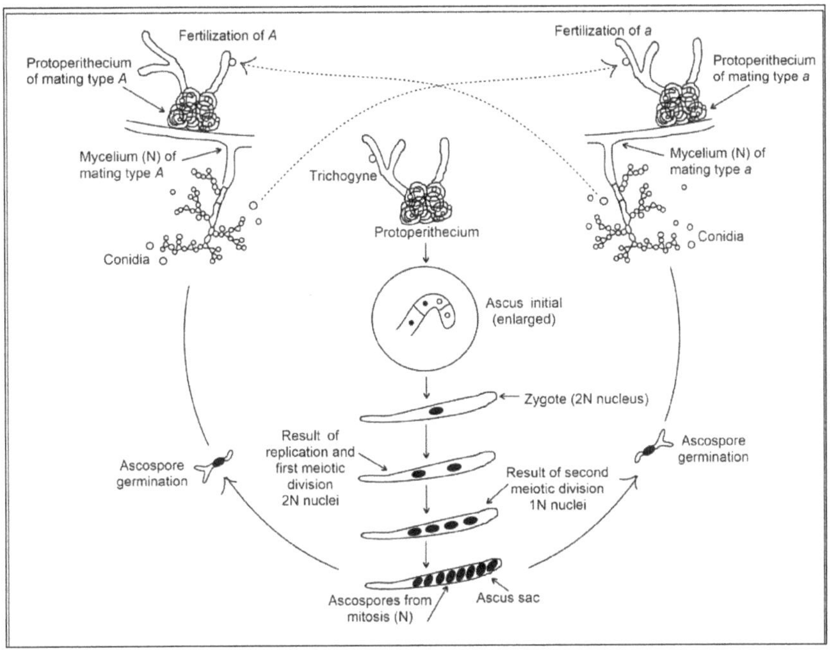

Neurospora crassa life cycle: The haploid mycelium reproduces asexually by two processes: (1) Simple proliferation of existing mycelium, and (2) formation of conidia (macro- and micro-) which can be dispersed and then germinate to produce new mycelium. In the sexual cycle, mating can only occur between individual strains of different mating type, A and a. Fertilization occurs by the passage of nuclei of conidia or mycelium of one mating type into the protoperithecia of the opposite mating type through the trichogyne. Fusion of the nuclei of opposite mating types occurs within the protoperithecium to form a zygote (2N) nucleus.

Neurospora crassa is a type of red bread mold of the phylum Ascomycota. N. crassa is used as a model organism because it is easy to grow and has a haploid life cycle: this makes genetic analysis simple, since recessive traits will show up in the offspring. Analysis of genetic recombination is facilitated by the ordered arrangement of the products of meiosis within a sac-like structure called an ascus (pl. asci). In its natural environment, N. crassa lives mainly in tropical and sub-tropical regions. It often can be found growing on dead plant matter after fires.

Neurospora was used by Edward Tatum and George Wells Beadle in the experiments for which they won the Nobel Prize in Physiology or Medicine in 1958. The results of these experiments led directly to the "one gene, one enzyme" hypothesis that specific genes code for specific proteins. This concept launched molecular biology. Sexual fruiting bodies (perithecia) can only be formed when two cells of different mating type come together. Like other Ascomycetes, N. crassa has two mating types that, in this case, are symbolized by A and a. There is no evident morphological difference between the A and a mating type strains. Both can form abundant protoperithecia, the female reproductive structure. Protoperithecia are formed most readily in the laboratory when growth occurs on solid (agar) synthetic medium with a relatively low source of nitrogen. Nitrogen starvation appears to be necessary for expression of genes involved in sexual development. The protoperithecium consists of an ascogonium, a coiled multicellular hypha that is enclosed in a knot-like aggregation of hyphae. A branched system of slender hyphae, called the trichogyne, extends from the tip of the ascogonium projecting beyond the sheathing hyphae into the air. The sexual cycle is initiated (i.e. fertilization occurs) when a cell, usually a conidium, of opposite mating type contacts a part of the trichogyne. Such contact can be followed by cell fusion leading to one or more nuclei from the fertilizing cell migrating down the trichogyne into the ascogonium. Since both A and a strains have the same sexual structures, neither strain can be regarded as exclusively male or female. However, as a recipient, the protoperithecium of both the A and a strains can be thought of as the female structure, and the fertilizing conidium can be thought of as the male participant.

The subsequent steps following fusion of A and a haploid cells have been outlined by Fincham and Day. and Wagner and Mitchell. After fusion of the cells, the further fusion of their nuclei is delayed. Instead, a nucleus from the fertilizing cell and a nucleus from the ascogonium become associated and begin to divide synchronously. The products of these nuclear divisions (still in pairs of unlike mating type, i.e. A/a) migrate into numerous ascogenous hyphae, which then begin to grow out of the ascogonium. Each of these ascogenous hyphae bends to form a hook (or crozier) at its tip and the A and a pair of haploid nuclei within the crozier divide synchronously. Next, septa form to divide the crozier into three cells. The central cell in the curve of the hook contains one A and one a nucleus. This binuclear cell initiates ascus formation and is called an "ascus-initial" cell. Next the two uninucleate cells on either side of the first ascus-forming cell fuse with each other to form a binucleate cell that can grow to form a further crozier that can then form its own ascus-initial cell. This process can then be repeated multiple times.

After formation of the ascus-initial cell, the A and a nuclei fuse with each other to form a diploid nucleus. This nucleus is the only diploid nucleus in the entire life cycle of N. crassa. The diploid nucleus has 14 chromosomes formed from the two fused haploid nuclei that had 7 chromosomes each. Formation of the diploid nucleus is immediately followed by meiosis. The two sequential divisions of meiosis lead to four haploid nuclei, two of the A mating type and two of the a mating type. One further mitotic division leads to four A and four a nucleus in each ascus. Meiosis is an essential part of the life cycle of all sexually reproducing organisms, and in its main features, meiosis in N. crassa seems typical of meiosis generally.

As the above events are occurring, the mycelial sheath that had enveloped the ascogonium develops as the wall of the perithecium becomes impregnated with melanin, and blackens. The mature perithecium has a flask-shaped structure.

A mature perithecium may contain as many as 300 asci, each derived from identical fusion diploid nuclei. Ordinarily, in nature, when the perithecia mature the ascospores are ejected rather violently into the air. These ascospores are heat resistant and, in the lab, require heating at 60 °C for 30 minutes to induce germination. For normal strains, the entire sexual cycle takes 10 to 15 days. In a mature ascus containing eight ascospores, pairs of adjacent spores are identical in genetic constitution, since the last division is mitotic, and since the ascospores are contained in the ascus sac that holds them in a definite order determined by the direction of nuclear segregations during meiosis. Since the four primary products are also arranged in sequence, a first division segregation pattern of genetic markers can be distinguished from a second division segregation pattern.

Benefit of Mating Type in N. Crassa

That mating in N. crassa can only occur between strains of different mating type suggests that some degree of outcrossing is favored by natural selection. In haploid multicellular fungi, such as N. crassa, meiosis occurring in the brief diploid stage is one of their most complex processes. The haploid multicellular vegetative stage, although physically much larger than the diploid stage, characteristically has a simple modular construction with little differentiation. In N. crassa, recessive mutations affecting the diploid stage of the life cycle are quite frequent in natural populations. These mutations, when homozygous in the diploid stage, often cause spores to have maturation defects or to produce barren fruiting bodies with few ascospores (sexual spores). The majority of these homozygous mutations cause abnormal meiosis (e.g. disturbed chromosome pairing or disturbed pachytene or diplotene). The number of genes affecting the diploid stage was estimated to be at least 435 (about 4% of the total number of 9,730 genes). Thus, outcrossing, promoted by the necessity for union of opposite mating types, likely provides the benefit of masking recessive mutations that would otherwise be deleterious to sexual spore formation.

Saccharomyces Cerevisiae

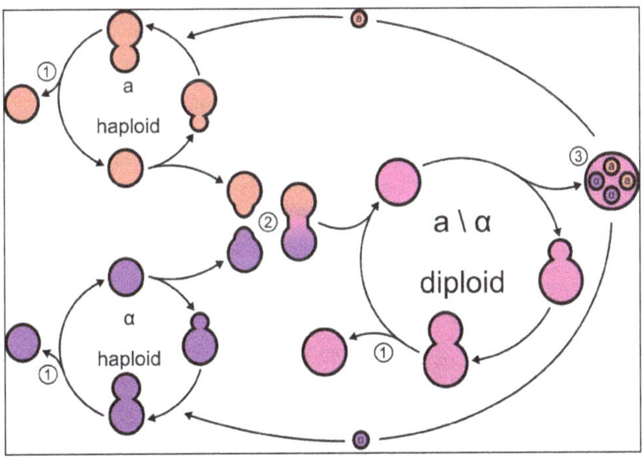

The yeast cell's life cycle: Budding, conjugation and spore.

Saccharomyces cerevisiae, brewer's and baker's yeast, is in the phylum Ascomycota. During vegetative growth that ordinarily occurs when nutrients are abundant, S. cerevisiae reproduces by mitosis as either haploid or diploid cells. However, when starved, diploid cells undergo meiosis to form haploid spores. Mating occurs when haploid cells of opposite mating type, MATa and MATα, come into contact. Ruderfer et al. pointed out that such contacts are frequent between closely related yeast cells for two reasons. The first is that cells of opposite mating type are present together in the same ascus, the sac that contains the tetrad of cells directly produced by a single meiosis, and these cells can mate with each other. The second reason is that haploid cells of one mating type, upon cell division, often produce cells of the opposite mating type with which they may mate.

Katz Ezov et al. presented evidence that in natural S. cerevisiae populations clonal reproduction and a type of "self-fertilization" (in the form of intratetrad mating) predominate. Ruderfer et al. analyzed the ancestry of natural S. cerevisiae strains and concluded that outcrossing occurs only about once every 50,000 cell divisions. Thus, although S. cerevisiae is heterothallic, it appears that, in nature, mating is most often between closely related yeast cells. The relative rarity in nature of meiotic events that result from outcrossing suggests that the possible long-term benefits of outcrossing (e.g. generation of genetic diversity) are unlikely to be sufficient for generally maintaining sex from one generation to the next. Instead, a short term benefit, such as meiotic recombinational repair of DNA damages caused by stressful conditions such as starvation, may be the key to the maintenance of sex in S. cerevisiae. Alternatively, recessive deleterious mutations accumulate during the diploid expansion phase, and are purged during selfing: this purging has been termed "genome renewal" and provides an advantage of sex that does not depend on outcrossing.

Candida Albicans

Candida albicans is a diploid fungus that grows both as a yeast and as a filament. C. albicans is the most common fungal pathogen in humans. It causes both debilitating mucosal infections and potentially life-threatening systemic infections. C. albicans has maintained an elaborate, but largely hidden, mating apparatus. Johnson suggested that mating strategies may allow C. albicans to survive in the hostile environment of a mammalian host. In order to mate C. albicans needs to switch from white to opaque cells. The latter are more efficient in mating and referred to as the mating competent cells of C. albicans. Mating in C. albicans is termed a parasexual cycle since meiosis is still not observed in C. albicans.

Mating Type

A picture of the mating type mechanism has begun to emerge from studies of particular fungi such as S. cerevisiae. The mating type genes are located in homeobox and encode enzymes for production of pheromones and pheromone receptors. Sexual reproduction thereby depends on pheromones produced from variant alleles of the same gene. Since sexual reproduction takes place in haploid organisms, it cannot proceed until complementary genes are provided by a suitable partner through cell or hyphal fusion. The number of mating types depends on the number of genes and the number of alleles for each.

Depending of the species, sexual reproduction takes place through gametes or hyphal fusion. When a receptor on one haploid detects a pheromone from a complementary mating type, it approaches the source through chemotropic growth or chemotactic movement if it is a gamete.

Mating in Basidiomycota

Some of the species within Basidiomycota have the most complex systems of sexual reproduction known among Fungi. In general for Fungi there are two main types of sexual reproduction: homothallism, when mating occurs within a single individual, or in other words each individual is self-fertile; and heterothallism, when hyphae from a single individual is self-sterile and needs to interact with another compatible individual for mating to take place. Additionally, mating compatibility in the Basidiomycota is further categorized into two types of mating systems: tetrapolar and bipolar.

Tetrapolar and Bipolar Mating System

Heterothallism is the most common mating system in Basidiomycota and in Agaricomycotina (the mushroom-forming Fungi) about 90% of the species are heterothallic. The tetrapolar type of mating system is ruled by two unlinked mating loci termed A and B (in Agaricomycotina) or b and a (in Ustilaginomycotina and Pucciniomycotina), both of which can be multiallelic. The combination of A and B (or b and a) alleles, termed mating type, determine the specificity or sexual identity of the individual harboring them. Only individuals with different specificities are compatible with each other and therefore able to start the mating event. A successful mating interaction begins with nuclear exchange and nuclear migration resulting in the formation of dikaryotic hyphae (containing separate haploid nuclei from both initial parents). Dikaryotic hyphae, under the appropriate environmental conditions will give rise to the fruiting body which contains the basidia – specialized cells in which sexual recombination via karyogamy and meiosis occurs. This dikaryotic condition in Basidiomycota is often maintained by a specialized hyphal structure called a clamp connection. The formation of clamp connections is regulated by both mating loci. Examples of tetrapolar organisms are the smuts Ustilago maydis and U. longissima, and the mushrooms Coprinopsis cinerea, Schizophyllum commune, Pleurotus djamor and Laccaria bicolor. It is believed that multi-allelic systems favor outcrossing in Basidiomycota. For example, in the case of U. maydis, which bears more than 25 b but only 2 a specificities, an individual has an approximately 50% chance to encounter a compatible mate in nature. However, species such as C. cinerea, which has more than 240 A and B specificities each, and S. commune, which has more than 339 A specificities and 64 B specificities, approach close to 100% chance of encountering a compatible partner in nature due to the huge number of mating types generated by these systems.

In contrast, bipolar mating systems are ruled by a single allelic mating locus, termed either A or b. In Agaricomycotina, bipolar organisms mostly have multiple alleles for their A mating locus; however, in Ustilaginomycotina and Pucciniomycotina, the b mating locus is predominantly diallelic, which reduces the occurrence of outcrossing within these species. Bipolarity likely arose via one of two potential routes: 1) during evolution the B or a locus lost functionality in determining specificity, as has occurred in the mushroom Coprinellus disseminatus; or 2) both mating loci have become physically linked such that they now act as a single locus; this has occurred in the smut plant pathogen U. hordei and in the human pathogen Cryptococcus neoformans. Virulence success in these two pathogens is highly associated with mating and their mating type locus. Other bipolar species include the white rot fungus Phanerochaete chrysosporium and the edible mushroom Pholiota nameko.

The A and B or b and a Mating Loci

In the B or a locus there are linked genes that code for pheromones and pheromone receptors. The pheromones are short polypeptides with conserved residues and the pheromone receptors belong to the G protein-coupled family of receptors located in the cell membrane; they sense different molecules (in this case the pheromones) outside and activate a specific pathway inside of the cell. Pheromone-receptor interaction occurs in a way that the pheromone from one individual interacts with the receptor from the partner and vice versa. The functions of these genes are to regulate reciprocal nuclear exchange, nuclear migration in both mates and ultimately clamp cell fusion. The first mating pheromone-receptor genes characterized were for U. maydis.

The A or b mating locus contains genes that code for two types of homeodomain transcription factor proteins, usually tightly linked, that are homologues to the Saccharomyces cerevisiae mating proteins MATα2 and MATa1. In Agaricomycotina the two types of homeodomain transcription factors are termed HD1 and HD2; so the HD1 and HD2 proteins from an individual interacts with the HD2 and HD1 proteins from the other partner, respectively, generating heterodimers able to activate the A transcriptional regulated pathway, which involves formation of clamp cells, coordinated nuclear division and septation.

Homothallism

Homothallic species may likely have evolved from heterothallic ancestors. In Basidiomycota homothallism is not very common and in Agaricomycotina it is estimated that only 10% of species have homothallic mating behavior. For example, one subspecies of the ectomycorrhizal Basidiomycete Sistotrema brinkmannii is homothallic, although other subspecies have maintained their ability to outcross. Also, a variety of the edible mushroom Agaricus bisporus, (A. bisporus var. eurotetrasporus) produces haploid self-fertile basidiospores. Additionally, in the human pathogen C. neoformans known to outcross under laboratory conditions, both mating types are not normally distributed in natural populations, with the α mating type much more commonly found (>99%), suggesting homothallism is the most prevalent mode of sexual reproduction in ´C. neoformans in nature. Finally, the fungus causing witches' broom in cacao, Moniliophthora perniciosa, has a primarily homothallic biology despite having A and B mating type-like genes in its genome.

Among the 250 known species of aspergilli, about 36% have an identified sexual state Among those Aspergillus species that exhibit a sexual cycle the overwhelming majority in nature are homothallic (self-fertilizing). Selfing in the homothallic fungus Aspergillus nidulans involves activation of the same mating pathways characteristic of sex in outcrossing species, i.e. self-fertilization does not bypass required pathways for outcrossing sex but instead requires activation of these pathways within a single individual. Fusion of haploid nuclei occurs within reproductive structures termed cleistothecia, in which the diploid zygote undergoes meiotic divisions to yield haploid ascospores.

SEXUAL SELECTION IN FUNGI

The significance of sexual selection, the component of natural selection associated with variation in mating success, is well established for the evolution of animals and plants, but not for the evolution of fungi. Even though fungi do not have separate sexes, most filamentous fungi mate in a hermaphroditic fashion, with distinct sex roles, that is, investment in large gametes (female role) and fertilization by other small gametes (male role). Fungi compete to fertilize, analogous to 'male-male' competition, whereas they can be selective when being fertilized, analogous to female choice. Mating types, which determine genetic compatibility among fungal gametes, are important for sexual selection in two respects. First, genes at the mating-type loci regulate different aspects of mating and thus can be subject to sexual selection. Second, for sexual selection, not only the two sexes (or sex roles) but also the mating types can form the classes, the members of which compete for access to members of the other class. This is significant if mating-type gene products are costly, thus signalling genetic quality according to Zahavi's handicap principle.

Fungal Sexual Selection Requirements

Most fungi can produce asexually and sexually. Currently, sexual selection has been studied to occur more predominantly in the Ascomycota and Basidiomycota phyla. Although different sexes are not present within fungi, sexual selection can act due to the presence of different sex roles as well as different mating types as most fungi are hermaphroditic. Sex roles can be distinguished in sexually producing filamentous fungi. For example, sexually producing ascomycetes can produce anisogamous gametes. The larger immobile gametes act as female gametes, while the smaller, motile gametes act as male gametes. Increased difference in the operational sex ratio (OSR) due to asymmetry between the sex roles leads to the production of more male gametes. In addition, variation within gamete quality which could affect offspring viability or fitness can also lead to differences in female/male gamete ratios.

In addition, sexual selection can occur within fungi if there is a limiting number of a certain type of gamete. The limiting gamete is typically the female gamete as they tend to be more costly to produce and invest more, energetically, in the zygote.

Most fungi have a haploid-diploid life cycle. Sexual selection is much more crucial in the diploid phase as the product of the phase immediately undergoes meiosis and can no longer be fertilized again.

Sexual Selection in Mushroom Fungi

Mushroom-forming fungi within Basidiomycetes produce sexually by the reciprocal migration of nuclei and have a male-biased OSR which aids in supporting that sexual selection is present within fungi. Although there are no traditional males present, there is variation between the mating types responsible for acting as the male or female sex role. Receiving mycelia act as the female gametes while the donating nucleus acts as the male gamete. Sexual selection might occur through male-male competition or by female choice. A study demonstrated that sexual selection does occur between the donating nucleus of the heterokaryon and the receiving homokaryon through the two nuclear types within the heterkaryon. These two nuclear types are in competition with each other to fertilize the homokaryon. Female choice between the nuclear types is also possible as there is a strong bias for one of the two nuclei. However, it is more supported that competition between the donating nuclei is more responsible as it has been shown that there is a variation in the success of various nuclei independent of the receiving mycelium. Some factors that may affect the nuclei's success include faster mitotic division, increased migration, and possibly a mechanism that suppresses mitotic division in the competition.

Most Basidiomycetes consist of two mating types, A and B, which are unlinked. However, the B locus codes for a larger amount of pheromones through subloci. Each subloci can produce many pheromones yet codes only one receptor resulting in a high redundancy of these pheromones. Although only one pheromone is necessary to fertilize, the presence of such a high amount increases the likelihood of that individual's pheromones to be received. This characteristic is hypothesized to have evolved as a result of sexual selection.

Male-male Competition in Fungi

Competition can occur within fungi as it does in plants and animals if an inequality in the ability to access one sex role or mating type is present. This inequality could result in increased competition for access to the other mating type. Competition can also occur if there is a variation in the quality of the gametes. Within fungi, there is competition to fertilize, which can be seen as the equivalent of male-male competition in plants and animals. Sexual selection in fungi aids in explaining certain characteristics including the high redundancy of pheromones in the B mating-type locus as well as strong pheromone signaling in yeasts.

Male gametes have the ability to reproduce asexually as asexual spores if they fail to reproduce sexually. Some fungal species are capable of producing male gametes of two different sizes. Throughout evolution, the smaller gametes have lost the ability to produce asexually in order to increase the likelihood of fertilizing a female gamete by decreasing size and increasing both the amount and motility of the male gamete. This is most likely a result of direct competition as a trade-off was made in order to increase the likelihood of fertilization.

Pheromone Signaling and the Presence of Male Choice

Pheromone signaling is used within fungi to either attract a mate or to assess the quality of that gamete; and tends to be more effective when occurring over small distances. Female gametes are typically the ones responsibly for producing pheromones in order to attract a mate. However, pheromones can also be released by conidia. Male choice can occur in populations where there is a low concentration of male gametes, allowing them to be selective in which female to fertilize. The female with the highest concentration of pheromone is usually chosen. In situations where fertilization may have to occur over long distances, water-soluble pheromones may be secreted as seen in the female gametes of an aquatic Chytridiomycetes. The same similarly occurs within an aquatic oomycete algae species; also, a variation in the pheromone production related to the male reaction has been observed. Variations within the pheromone production as well as the response have been studied and found are affected by environmental conditions as well as developmental differences between the fungi. Although not proven, the production of pheromones may lead to Fisherian runaway selection in which the production of pheromone increases due to the increase in its preference throughout generations.

In addition to pheromone signaling being used as a method in finding a mate, it also appears to be utilized as a method to assess mate quality. Pheromones are costly to produce due to post-translational modification and therefore may be subject to the handicap principle in which an organisms cannot fake its fitness by producing pheromones. In the isogamous budding yeast, Saccharomyces cerevisiae, pheromones are used to distinguish between different mating types and show a preference for higher concentrations of pheromones along with a rejection of lower concentrations.

Possible Polyandry and Female Choice

Post-copulatory mechanisms may also be present within fungi through polyandry in which zygote-level sexual selection might occur. Within multicellular ascomycete fungi, a haploid mycelium produces a fruiting body which in turn produces many offspring that are also haploid. Each fruiting body has the potential to be fertilized by more than one male gamete. Laboratory experiments have shown that multiple matings are possible and the female has the ability to selectively abort fruiting bodies that have been inappropriately fertilized by a closely related yet incompatible species.

Possible Female-female Competition in Fungi

A study has shown that female-biased genes evolve must faster than male-biased genes demonstrated by a rapid protein-level evolution. In most other eukaryotes, male-biased genes demonstrate a faster evolution. This may be due to specific reproductive traits within fungi, possibly allowing female-female competition to occur rather than male-male competition. This is also supported the passive nature of conidia during mating. Due to the presence of pheromones as well as the dispersal of male gametes, two or more female gametes may be attracted to one conidium, or male gamete. The presence of karyogamy further supports the possibility of female-female competition. Within N. crassa, the haploid mycelium undergoes growth as vegetative tissue prior to entering the mating cycle. This vegetative tissue can be used as a source of fertilization and can fuse with the trichogyne. This causes the female gamete to no longer be subject to male-male competition and further fertilization by a conidium. The female-biased genes that were studied and seen to evolve more rapidly than male-biased genes did so regardless of the mechanism of fertilization. Fertilization occurring by the fusion of a female nucleus and male conidium nucleus or between a female nucleus or nucleus from vegetative tissue did not alter this rate of evolution.

FUNGAL SPORES

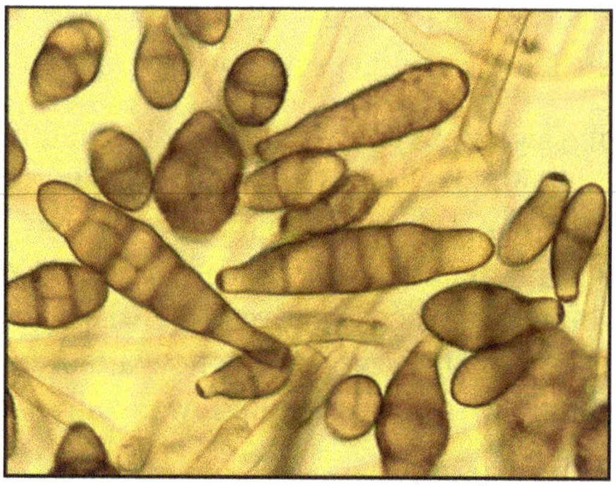

Spores of Alternaria.

Fungal spores are microscopic reproductive structures in fungi. Their function is similar to that of seeds in plants. Fungi, which include moulds, yeasts, mushrooms and toadstools, are an important component of the ecosystem. Fungi cause damage worth billions of dollars annually by causing diseases of animals and crops. While they also cause damage through biodeterioration and their spores pollute indoor environment, they help in the recycling of minerals and carbon by the decomposition of organic debris and waste. Every year trillions of tons of garbage are recycled worldwide by fungi.

Fungi used to be classified as plants. However, at the cell level, fungi are more like animals than they are like plants. Other characteristics that distinguish fungi from plants is that fungi cannot synthesize their own food like plants do and they have no roots.

Fungi spread either by forming reproductive spores that are carried on wind and rain or by growing and extending their hyphae. Hyphae are thread-like, strands of cells and grow as new cells form at the tips. A network of branching hyphae is called mycelium (plural mycelia).

Fungal Spore Release

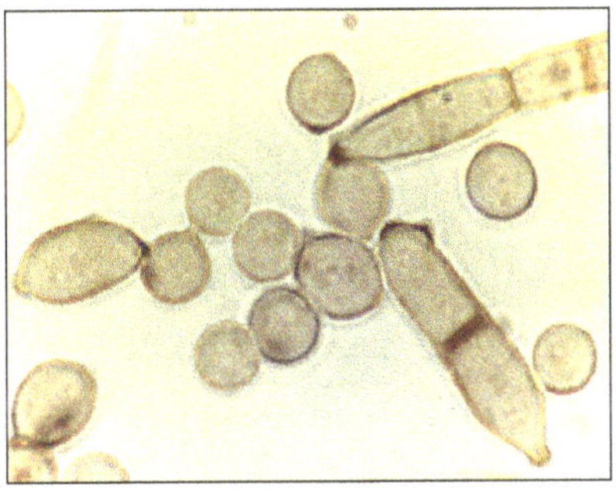

Cladosporium spores.

Fungal spores vary greatly in size, shape, color and method of release. Spores produced on raised structures are passively released by air movement. Many of these spores are delicately attached such that slight air movements or any disturbance of the substrate causes them to become airborne. The concentration of dry airborne spores e.g., Cladosporium can increase at the start of rainfall. Rapid air movement in advance of splashes and vibration can blow or tap spores into the air.

Spores of some fungi are released seasonally rather than throughout the year. Seasonal release of fungal plant pathogen spores coincides with a particular growth stage of the host plant. Ascospores and basidiospores are forcibly discharged into the air after wetting by rain or dew. The water creates pressure to force the ascospores from the ascii.

Airborne Fungal Spores in Houses

Several fungal spore types are frequently detected in the indoor environment. Since most of these spores originate from outdoors their counts are at levels generally lower than those detected in the outdoor samples. Some spores such as those of Penicillium/Aspergillus group are the most prevalent types indoors, usually exceeding the absolute levels and relative percentages of these spores outdoors. Cladosporium, Alternaria, Ascospores, and Basidiospores are the most prevalent spore types outdoors.

Moisture-indicator fungi such as Chaetomium, Stachybotrys, and Ulocladium species are very common in water damaged buildings and very rarely are they found in outdoor air samples.

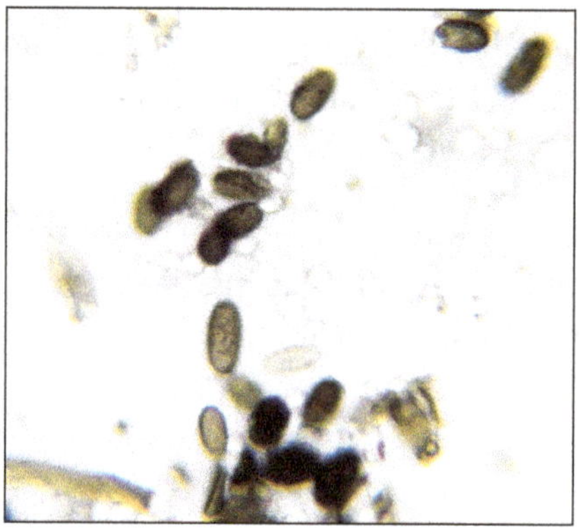

Stachybotrys spores.

Health effects of Fungal Spores

In spring, summer and fall in temperate climates outdoor spore counts may be extremely high. Inhalation of airborne fungal spores has been implicated in several ailments in humans. Airborne fungal spores contain allergens which can trigger a range of respiratory issues in humans such as allergies, asthma, and pathogenic infections of the respiratory tract. It is estimated that around 20% of the general population is atopic and easily get symptoms from normal fungal spores concentrations. Symptoms may range from mild allergies to severe asthma. A number of fungal spore types share similar allergens which means that those who are allergic to moulds are likely to react to multiple types of spores. Due to the fact that the different fungi vary in the release times of their spores, some people can therefore suffer most of the year.

Airborne fungal spores are also important agents of plant disease, and the means for dissemination of many common saprotrophic (saprophytic) fungi.

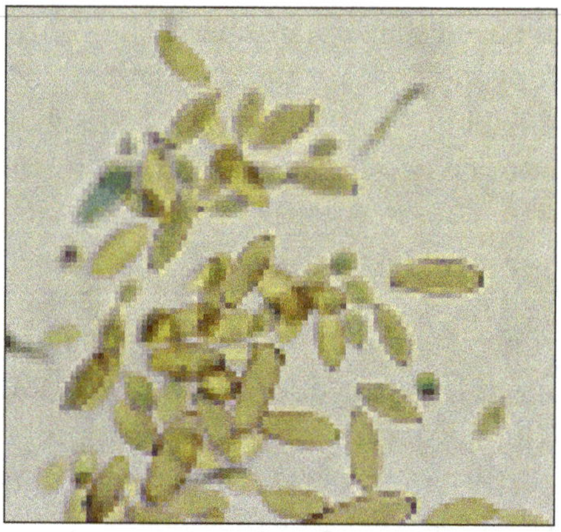

Cladosporium spores.

The most well known allergy and asthma causing fungal spore types worldwide are Alternaria, Cladosporium, and Aspergillus/Penicillium. Allergic reactions to each spore type differ between persons and the allergens vary in the severity of the allergic reaction they induce. More people are allergic to Alternaria than Cladosporium, for example, even though the latter is much more common in the air. Alternaria also produces more strongly positive reactions while Cladosporium tends to produce a milder allergic reaction. However, Cladosporium, and in particular Cladosporium herbarum, is often the major contributor to air-spora and due to its high concentrations is therefore a major cause of inhalant allergy and allergic asthma in humans.

Outdoor Fungal Spore Seasons

Fungal spores are released and distributed in a variety of ways and have specific climatic requirements. Some types favour warm, dry weather and are wind-dispersed and the numbers of these can be very high in the outdoor air. Examples of these 'dry air' spores include Cladosporium, Alternaria, Epicoccum, Drechslera, Pithomyces, and Curvularia, and smut spores. Other types need certain temperature and moisture thresholds to be reached before they are released, so a warm, humid period can also produce a high spore risk. Examples of 'wet air' spores include basidiospores and ascospores which are often abundant on mild, humid nights in summer.

Fungal spores are mainly seasonal in their release but many types can be found in the air all year round typically peaking in the summer or late fall.

Spring: Fungal spores are generally low in number but the risk can rise to moderate in the late Spring if conditions are favourable.

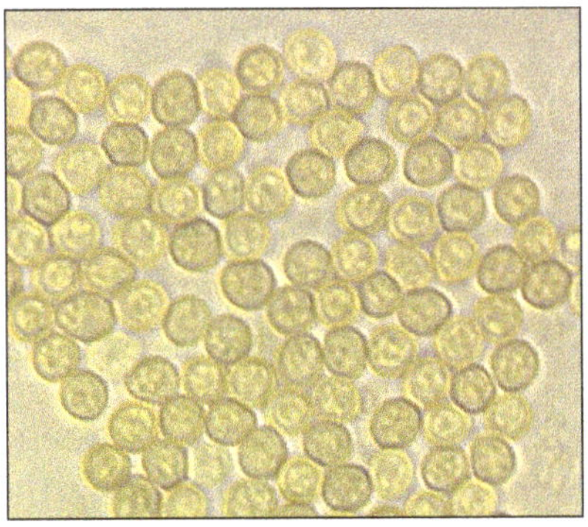

Trichoderma spores.

Summer to early fall: The spore risk starts to rise in mid-June with the increase in temperatures and in late June/early July the risk starts to peak and continues until late September. Typically, during this season, there will be the 'dry air' spore types of Alternaria & Cladosporium and some types of Penicillium and Aspergillus and others by day and then the 'wet air' spores that respond to the dew during the night. Therefore, there are a wide range of types that can affect people during each 24 hour period and many of the types have shared allergens that cross-react. In addition, rainfall will help the production of spores which are then released later after the rain (if it's heavy) or during light showers or drizzly rain. Very windy weather tends to dilute the spores in the air and unusually cold, dry weather will lead to a lower spore risk.

Mid-fall: Many of the main allergenic types are going into decline but there are still plenty of spores around and the risk can be moderate or even high in October on warmer, humid days. In November, the risk decreases to low by the end of the month unless it is very mild and damp. The lowest risk is on dry, cold frosty days.

Winter: Winter can be okay for some sufferers of fungal spores but for those sensitive to Penicillium and Aspergillus there will be some risk, particularly in January and early February.

Sporangiospore

Sporangiospores are spores that are produced in a sporangium (plural: sporangia). A sporangium in fungi (but not mosses and some other organisms) is simply a cell containing spores.

Asexual sporangia are commonly produced by the Chytridiomycota and Zygomycota. Although some similar structures, such as asci, are produced by members of the Dikarya they are never called sporangia and instead have their own special terms.

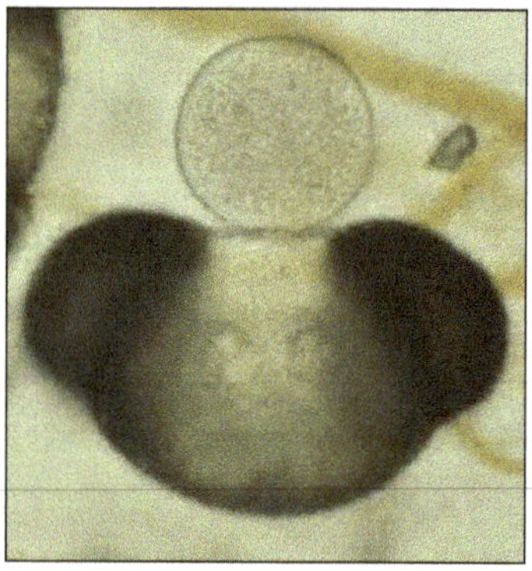

The picture above is the same one used in our discussion of sexual reproduction, except that it was photographed using bright-field rather than phase-contrast optics. It depicts a species of Rhizophidium, a member of the Chytridiomycota, growing on the upper surface of a grain of pine pollen floating in a pond. This individual is typical of many chytrids in its simplicity; the sporangium is essentially the whole fungus. It started to grow on the pollen grain when a spore landed there and became attached by rhizoids, seen as faint threads extending into the pollen grain. The developing spore swelled to form the spherical sporangium and its nucleus divided numerous times by mitosis until the sporangium became filled with nuclei. Each nucleus then became surrounded by a spore wall. At the time the photograph was taken the sporangium was filled with spores, faintly seen in the picture as an internal granularity. If it had been allowed to develop further the sporangium would have developed an exit tube through which the sporangiospores could escape. The sporangiospores of Rhizophidium have whip-like flagella, allowing them to swim off in search of new sources of food. Because of their motility the sporangiospores of the Chytridiomycota require liquid environments, or at least water films. Because they are asexual, these sporangia can develop quickly and produce large numbers of offspring, each of which can swim off and colonize a new pollen grain, soon producing even more progeny. Some species of Chytridiomycota produce more than one sporangium, and these remain connected by thin hypha-like threads.

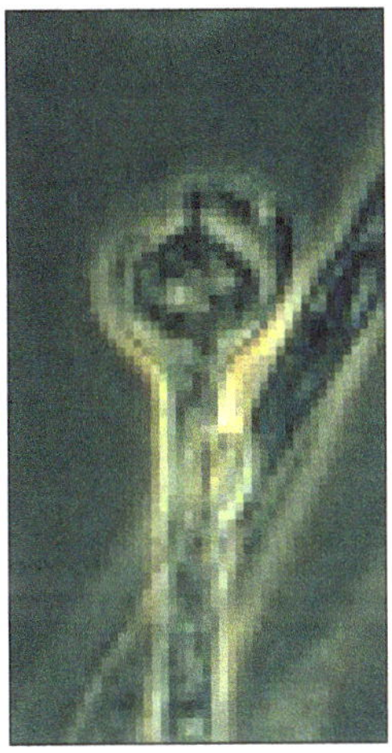

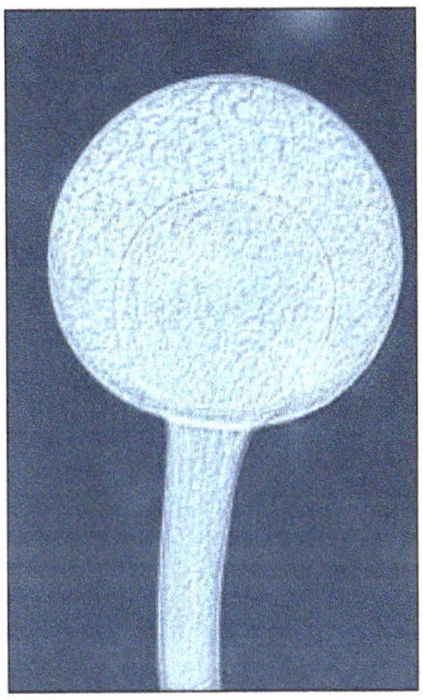

The Zygomycota are terrestrial organisms producing non-motile sporangiospores. The sporangia in this group are usually at the tips of specialized hyphae called sporangiophores. Probably the most difficult feature of the zygomycotan sporangium for most students to understand is the bulb-like structure called a columella. The columella is simply the end of the sporangiophore, enlarged so that it extends up into the sporangium. The picture of Rhizopus domesticus at near-left shows these relationships fairly well. The columella can be seen as a spherical body inside the large spherical sporangium. If you examine the photo carefully you will see where the columella connects to the sporangiophore on the right side. The spores inside the sporangium surround the columella but are never inside it. The photo at far-left shows a sporangiophore of Absidia spinosa. Here the sporangium has ruptured and released all of its spores. What is left is the sporangiophore topped by a columella, with the remnants of the sporangial wall remaining as a cup. The columella of Absidia spinosa, and that of some other Zygomycota as well, is characterized by the presence of an apophysis, a small pointed extension on its upper surface. It is these "naked" columellae, free of the sporangium and spores, that cause beginners so much confusion. They give the appearance of being immature sporangia, and may be interpreted as such if intact sporangia are not present. Of course some Zygomycota produce sporangia without a columella, but these are usually fairly easy to recognize.

Zygospore

A zygospore is a diploid reproductive stage in the life cycle of many fungi and protists. Zygospores are created by the nuclear fusion of haploid cells. In fungi, zygospores are formed in zygosporangia after the fusion of specialized budding structures, from mycelia of the same (in homothallic fungi) or different mating types (in heterothallic fungi), and may be chlamydospores. In many eukaryotic algae, including many species of the Chlorophyta, zygospores are formed by the fusion of unicellular gametes of different mating types.

A zygospore remains dormant while it waits for environmental cues, such as light, moisture, heat, or chemicals secreted by plants. When the environment is favorable, the zygospore germinates, meiosis occurs, and haploid vegetative cells are released.

In fungi, a sporangium is produced at the end of a sporangiophore that sheds spores. A fungus that forms zygospores is called a zygomycete, indicating that the class is characterized by this evolutionary development.

Germination of Zygospore

The ways of zygospore germination in lower fungi:

- The diploid contents of the zygospore (A) undergo meiosis but meiospores are not formed. The zygospore containing the haploid protoplast then directly germinates to give rise to new haplomycelium. (B) Monoblepharis is an example.

  Rarely as in Allomyces, the diploid zygospore (C) germinates by putting out a germ tube (D) which grows and branches to form the diplomycelium. No meiosis takes place prior to or during germination of the zygospore.

- Prior to germination, the diploid unclei undergo zygotic meiosis. The resultant haploid nuclei are fashioned into meiospores. The latter are extruded either by the rupture of zygospore wall into a sessile vesicle (E, Albugo Candida) or the zygospore forms a short exit tube which terminates in a vesicle (F, Pythium perioplocum).

  With the bursting of the vesicle wall, the meiospores are liberated. They swim about for a while, retract their flagella and encyst. The encysted meiospore germinates by a germ tube.

- In some other species, the diploid contents of the zygospore, prior to germination, undergo meiosis. Upon germination, the haploid contents grow into a hypha which grows to a limited extent. It is called the promycelium. The latter bears a terminal sporangium or germ sporangium.

  The haploid contents of the germ sporangium undergo cleavage to form either motile (Plasmopara viticola, G), or non-motile (Mucor, H) germ spores. The former are called meiozoospores and the latter meioaplanospores. On liberation, each germ spore germinates under suitable conditions to form the haplomycehum.

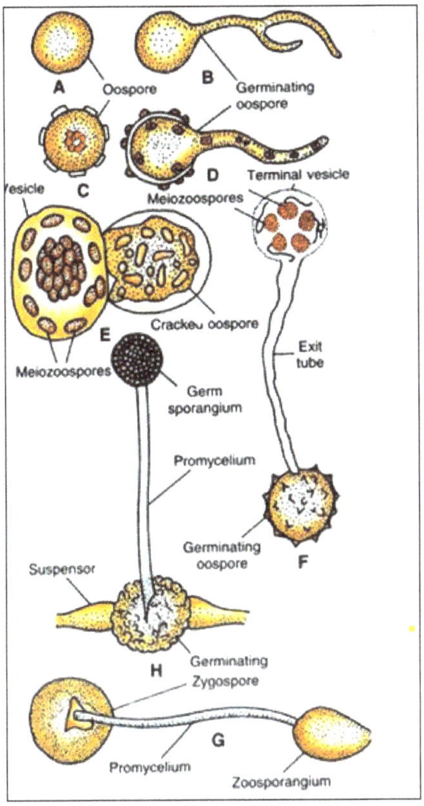

Lower fungi: Germination of zygospore. Explanation in the text. (A-B, after smith C-D, after woronin. E, after de Bary. F after drechsler and G after avans.

Ascospore

Ascospores are microscopic spores produced by fungal species belonging to the Ascomycota. With over 64 000 species in this division, it is apparent that the term Ascospores covers a wide range and variety of spores. However broad the term may be, a single airborne spore, upon reaching a suitable damp, humid environment, can easily develop into a multicellular fungal growth, more commonly known as a mold. Apart from being aesthetically displeasing, the mold will release more spores into the surrounding air, potentially leading to health issues for any inhabitants.

Knowledge is power when it comes to finding indoor type spores and removing them permanently. If you suspect you have found a potential mold growth, don't wait to call a professional team of mold specialists or testers for a solution.

These sexual spores are formed in capsules known as asci. They can be found almost everywhere and can belong to one of more than 3000 genera that include Chaetomium, Aspergillus, Penicillium,Peziza, Ascotricha and many more.

Taphrina padi with Ascospores.

However, these molds mainly grow on ornamental trees, fruits, crop plants and timber, while others are associated with insects and even fish. Of special interest are the Cordyceps and Opiocordyceps genera, several species of which have evolved to target a specific host species, usually a species of ant or wasp. Once infected, the arthropod will exhibit behavioural changes, all to the benefit of the fungus. Thanks to these apparent "mind controlling" properties, the Cordyceps genus has been popularized in science fiction. Thankfully, they live in remote tropical forests and have not evolved to parasite on humans (yet).

Moving on, species of Ascomycota and therefore their spores can be categorized in terms of health significance as well. Those not representing any indoor air quality significance are the ones that are generally used for morphological identification.

In fact, the life cycle of most Ascomycota species can be divided into two distinct reproductive phases–sexual and asexual. While the former phase involves the formation of Ascospores, the topic of conversation, the latter involves asexual spores called conidiospores, which are produced on specialized stalks (conidiophores).

Formation of Ascospores

Depending on the species, a large variety of conditions can help form Ascospores in the sac-like structure of an ascus. The spores are formed after the fusion of male and female structures, the antheridium and the ascogonium. Ascogenous hyphae will grow out of the fertilized ascogonium, and after several meiotic and mitotic divisions, the spores mature. In most cases, an ascus holds 8 spores. However, the number can vary from one to several thousand.

These microscopic spores of can be found developing on dead, decaying and infected plants during the winter. During the spring, warm rains trigger the release of spores into the air and that's exactly how the infections develop as the number of Ascospores grows.

Moreover, these molds cause diseases like powdery mildews, ear rot of corn, blue stains, fruit rot of cereals and many more.

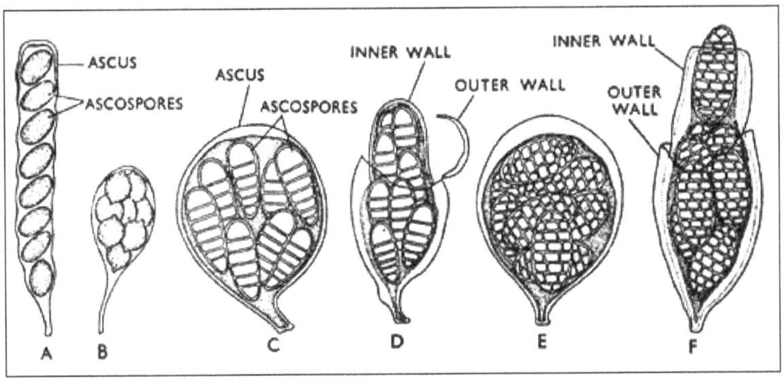

Nature of unitunicate and bitunicate asci. A-B: Unitunicate asci. C-F: Bitunicate asci. C and E: Intact Bitunicate asci. D: Expanding inner wall splitting the outer wall. F: Escape of ascospores through a pore at the expanded inner wall. Nature of asci and ascopores.

Ascospores Mold

Ascospores are a phase in the life cycle of many fungi belonging to the Ascomycotadivision. With thousands of known species, these fungi are highly variable and can grow as budding single cells such as yeasts, or can grow into a macroscopic mycelium, more commonly known as a mold. These molds can come in many shapes, sizes and colors, with varying transparency/opaqueness.

In the house, you can generally find them everywhere, from bathroom tiles to the ceiling, on floors, walls and doors.

However, although the complete list of their habitats is practically endless, the following list suggests areas to keep an eye on:

- Windows.
- Upholstered furniture.
- Rugs and mattresses.
- Walls and cabinets.
- Airtight closets.

In short, these molds can grow relatively well in damp, enclosed, wet or air sealed surfaces or materials.

Basidiospore

A basidiospore is a reproductive spore produced by Basidiomycete fungi, a grouping that includes mushrooms, shelf fungi, rusts, and smuts. Basidiospores typically each contain one haploid nucleus that is the product of meiosis, and they are produced by specialized fungal cells called basidia. Typically, four basidiospores develop on appendages from each basidium, out of these 2 are of one strain and other 2 of its opposite strain. In gills under a cap of one common species, there exist millions of basidia. Some gilled mushrooms in the order Agaricales have the ability to release billions of spores. The puffball fungus Calvatia gigantea has been calculated to produce about five trillion basidiospores. Most basidiospores are forcibly discharged, and are thus considered ballistospores. These spores serve as the main air dispersal units for the fungi. The spores are released during periods of high humidity and generally have a night-time or pre-dawn peak concentration in the atmosphere.

When basidiospores encounter a favorable substrate, they may germinate, typically by forming hyphae. These hyphae grow outward from the original spore, forming an expanding circle of mycelium. The circular shape of a fungal colony explains the formation of fairy rings, and also the circular lesions of skin-infecting fungi that cause ringworm. Some basidiospores germinate repetitively by forming small spores instead of hyphae.

General Structure and Shape

Basidiospores are generally characterized by an attachment peg (called a hilar appendage) on its surface. This is where the spore was attached to the basidium. The hilar appendage is quite prominent in some basidiospores, but less evident in others. An apical germ pore may also be present. Many basidiospores have an asymmetric shape due to their development on the basidium. Basidiospores are typically single-celled (without septa), and typically range from spherical to oval to oblong, to ellipsoid or cylindrical. The surface of the spore can be fairly smooth, or it can be ornamented. The color of the spore print is usually found in the spore wall, although in rare instances – like the yellow spores of Clavaria helicoides – the cytoplasm is responsible for the spore color.

Aeciospore

Aeciospore (aecidiospore) is a spore produced by certain fungi of the order Uredinales. Chains of aeciospores are produced in an aecium which develops in the tissues of the host plant.

Ascus

An ascus is the sexual spore-bearing cell produced in ascomycete fungi. Asci usually contain eight ascospores, produced by meiosis followed, in most species, by a mitotic cell division. However, asci in some genera or species can occur in numbers of one (e.g. Monosporascus cannonballus), two, four, or multiples of four. In a few cases, the ascospores can bud off conidia that may fill the asci (e.g. Tympanis) with hundreds of conidia, or the ascospores may fragment, e.g. some Cordyceps, also filling the asci with smaller cells. Ascospores are nonmotile, usually single celled, but not infrequently may be coenocytic (lacking a septum), and in some cases coenocytic in multiple planes. Mitotic divisions within the developing spores populate each resulting cell in septate ascospores with nuclei. The term ocular chamber, or oculus, refers to the epiplasm (the portion of cytoplasm not used in ascospore formation) that is surrounded by the "bourrelet" (the thickened tissue near the top of the ascus).

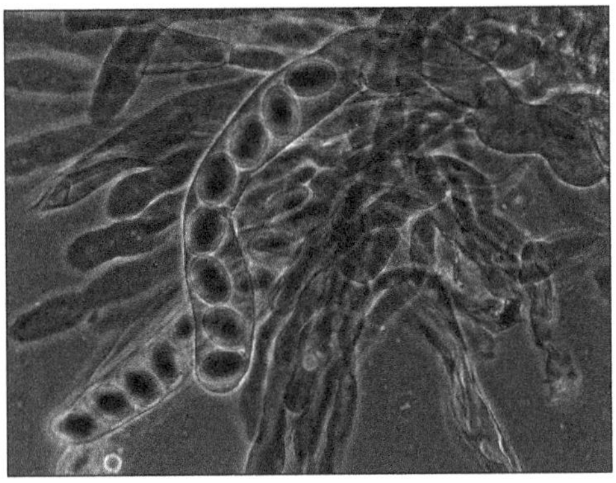

Asci of Morchella elata, Phase contrast.

In many cases the asci are formed in a regular layer, the hymenium, in a fruiting body which is visible to the naked eye, here called an ascocarp or ascoma. In other cases, such as single-celled yeasts, no such structures are found. In rare cases asci of some genera can regularly develop inside older discharged asci one after another, e.g. Dipodascus.

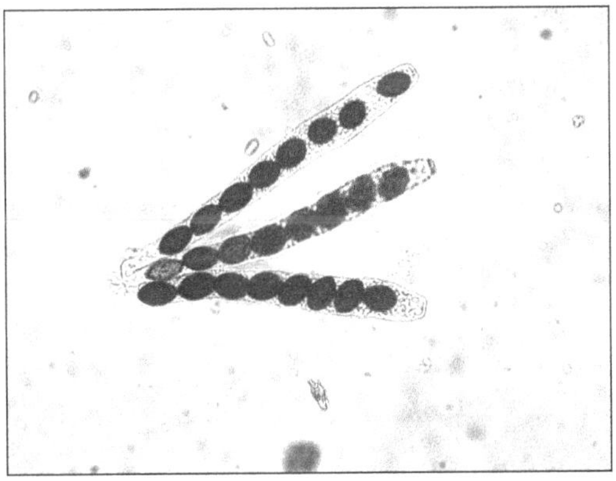

There are 8 ascospores in each ascus of Sordaria fimicola.

Asci normally release their spores by bursting at the tip, but they may also digest themselves passively releasing the ascospores either in a liquid or as a dry powder. Discharging asci usually have a specially differentiated tip, either a pore or an operculum. In some hymenium forming genera, when one ascus bursts, it can trigger the bursting of many other asci in the ascocarp resulting in a massive discharge visible as a cloud of spores – the phenomenon called "puffing". This is an example of positive feedback. A faint hissing sound can also be heard for species of Peziza and other cup fungi.

Asci, notably those of Neurospora crassa, have been used in laboratories for studying the process of meiosis, because the four cells produced by meiosis line up in regular order. By modifying genes coding for spore color and nutritional requirements, the biologist can study crossing over and other phenomena.

Asci of most Pezizomycotina develop after the formation of croziers at their base. The croziers help maintain a brief dikaryon. The compatible nuclei of the dikaryon merge forming a diploid nucleus that then undergoes meiosis and ultimately internal ascospore formation. Members of the Taphrinomycotina and Saccharomycotina do not form croziers.

Classification

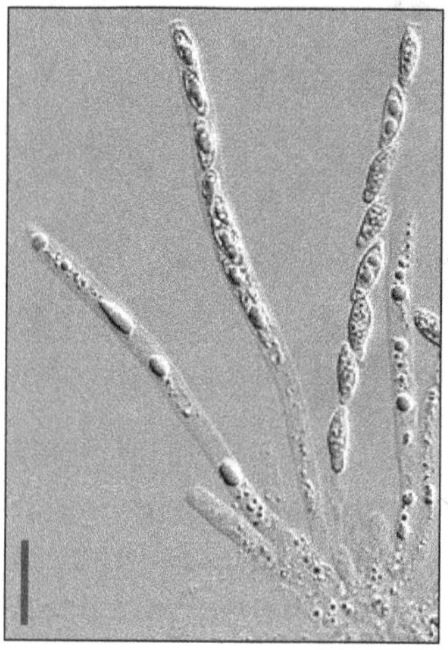

Asci of Hypomyces chrysospermus (they are unitunicate-inoperculate).

The form of the ascus, the capsule which contains the sexual spores, is important for classification of the Ascomycota. There are four basic types of ascus:

- A unitunicate-operculate ascus has a "lid", the Operculum, which breaks open when the spores are mature and allows the spores to escape. Unitunicate-operculate asci only occur in those ascocarps which have apothecia, for instance the morels. 'Unitunicate' means 'single-walled'.
- Instead of an operculum, a unitunicate-inoperculate ascus has an elastic ring that functions like a pressure valve. Once mature the elastic ring briefly expands and lets the spores shoot out. This type appears both in apothecia and in perithecia; an example is the illustrated Hypomyces chrysospermus.

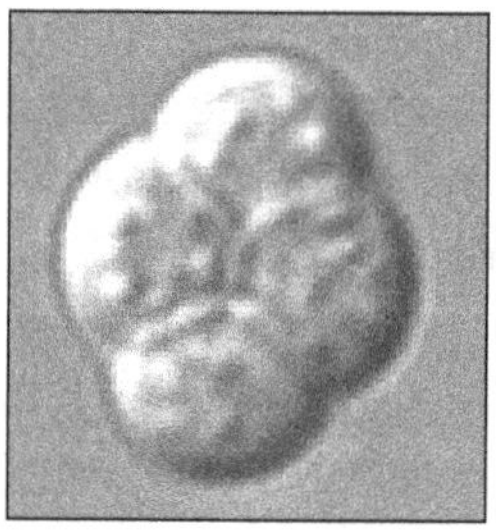

Ascus of Saccharomyces cerevisiae containing a tetrad of four spores.

- A bitunicate ascus is enclosed in a double wall. This consists of a thin, brittle outer shell and a thick elastic inner wall. When the spores are mature, the shell splits open so that the inner wall can take up water. As a consequence this begins to extend with its spores until it protrudes above the rest of the ascocarp so that the spores can escape into free air without being obstructed by the bulk of the fruiting body. Bitunicate asci occur only in pseudothecia and are found only in the classes Dothideomycetes and Chaetothyriomycetes (which were formerly united in the old class Loculoascomycetes). Examples: Venturia inaequalis (apple scab) and Guignardia aesculi (Brown Leaf Mold of Horse Chestnut).

- Prototunicate asci are mostly spherical in shape and have no mechanism for forcible dispersal. The mature ascus wall dissolves allowing the spores to escape, or it is broken open by other influences, such as animals. Asci of this type can be found both in perithecia and in cleistothecia, for instance with Dutch elm disease (Ophiostoma). This is something of a catch-all term for cases which do not fit into the other three ascus types, and they probably belong to several independent groups which evolved separately from unitunicate asci.

Oospore

An oospore is a thick-walled sexual spore that develops from a fertilized oosphere in some algae, fungi, and Oomycetes. They are believed to have evolved either through the fusion of two species or the chemically-induced stimulation of mycelia, leading to oospore formation.

In Oomycetes, oospores can also result from asexual reproduction, by apomixis. These are found in Fungi as the sexual spores; these help in the sexual reproduction of Fungi. These are non motile spore these are also haploid spores these are site of meiosis and karyogamy in oomycetes .

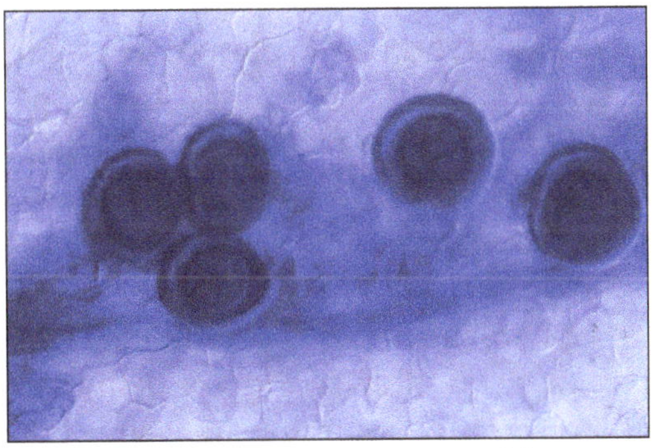

Oospores of Hyaloperonospora parasitica, agent of the downy mildew (in the middle).

Nuclear fusion in the oosphere of watermolds followed by the accretion of a thick wall and accompanied by certain internal cytoplasmic and reserve material adjustments creates the oospore or mature zygote in species of Saprolegniaceae. Ultrastructural analyses by Beakes and Gay, however, suggest that the usual concept of an oospore must be modified to some degree.

Oospore maturation progresses through four phases, and the definition of this structure is to be found by tracing events in that progression. The oosphere stage (two phases) begins with the onset of cytoplasmic furrowing and ends with the acquisition of an outer wall and the initiation of an inner one. At this point, some 8-15 hours after cleavage begins, fertilization takes place. The inner wall then is completed, the central ooplast vacuole develops, and the cell is now properly termed an oospore. The fourth phase is a period of unchanging internal organization – after-ripening, terminating in germination.

Two salient points regarding oospore structure emerge from the foregoing sequential definition offered by Beakes and Gay and from earlier studies by Howard and Howard and Moore. The first point is that the oospore wall is not a single layer. De Bary maintained that spores had three wall layers: the exospore (derived from condensed periplasm), the epispore, and the endospore (secondary wall layers deposited during zygote maturation). The outer layer consists of the zygote membrane together with the oosphere membrane and any accompanying deposits. The endospore or inner wall is thick and usually stratified. Electron microscopy applied to the oospore investiture shows it to be much more complex than ordinary reveals. Beakes and Gay have demonstrated that in Saprolegnia furcata, at least, the oospore has an outer wall, and an inner one consisting of an inner and an outer layer separated by an electron dense accretion.

The second point concerning oospore structure also emerges from ultrastructural studies. Mature oospores of Saprolegnia terrestris contain an ooplast. Howard and Moore reported that this structure is in fact a large vacuole, but which has been traditionally defined and illustrated as the protoplasm of the zygote. According to their interpretation of the oospore, the cytoplasm proper (containing the organelles and reserve globules) surrounds or partially encompasses the ooplast. Beakes and Gay also recognized the ooplast ultrastructurally.

Using Saprolegnia diclina, J. Fletcherhas explored very thoroughly the appearance and timing of events in the process by which oospheres mature into oospores. He found that the occurrence of the stages in maturation was quite variable, ranging from one-quarter to twice the calculated mean times. The extremes in variation in time could occur among oospores in a single oogonium as well among those from different oogonia. The mean time elapsed between the delimitation of the oogonium and the appearance of identifiable oosphere initials was 9 hours 25 minutes; discrete oospheres appeared 10 hours and 55 minutes after oogonium delimitation. On the average, 12 hours and 46 minutes after discrete oospheres were visible in the oogonia of S. diclina, an identifiable oospore wall was produced. The central ooplast was formed at 45 hours 52 minutes (mean time) after condensed, spherical oospheres were present, and the reserve bodies were aligned peripherally - signaling oospore maturation - at 66 hours and 47 minutes after the oosphere was completed. There are several recognizable stages in the maturation process in S. diclina. Following the condensation of the oosphere initial into the oosphere an oospore wall appears. Subsequently, the granular protoplast separates from the wall leaving a clear zone. In time, this clear, peripheral area seems to be occupied, at least in part, by the reserve globules that develop as maturation reaches the climax. The ooplast becomes visible in the maturing zygote first as a number of well-defined globose structures; these eventually coalesce into one body, the ooplast proper. The final stage in maturation is the peripheral alignment of the small reserve globules.

The study by J. Fletcher should be repeated with other species to determine if the sequential events and the timing of them are of general occurrence among the watermolds or only an unusual instance in one isolate. Such explorations might well begin with the dioecious Achlyas, taking into account the procedure developed by Rozek and Timberlake to induce synchronous oospore production in Achlya ambisexualis. Encysted spores from the oogonial (female) thallus and from the antheridial (male) thallus were mixed, and the resulting mycelial growth analyzed for oogonium and oospore production. The number of oogonia produced was found to be directly related to the proportion (in the original inoculum) of "female" spore cysts. A ratio of seven oogonial thallus spore cysts to three from the antheridial thallus was optimum for the formation of oogonia with oospores.

The Types of Oospores

Classically, three types of oospores have been recognized based upon what is ordinarily thought of as the arrangement of oil globules in the cytoplasm. Because these deposits are chemically more complex than lipid alone, Dick substituted the term reserve globule; we shall refer to this material by this latter designation or simply as a refractive material.

In view of the study by Howard and Moore it is necessary to modify the more familiar definitions of oospore types to take into account the ooplast. A centric oospore possesses peripherally in the layer of cytoplasm that surrounds the ooplast one or two complete concentric spheres of small reserve globules. In a subcentric oospore the cytoplasm only partially encompasses the ooplast, and contains two or three peripheral partial spheres of oil reserve globules in one area, or, in some cases, a full peripheral sphere of bodies in addition. In optical section, such an oospore generally is seen to have two or three rows of refractive bodies at one side (gathered in a lunate configuration), and only one row on the opposing side. In some cases, particularly where the oospore is ellipsoidal and is viewed in optical section, two or three "lines" of globules may be present at each end of the cell, with a single "layer" connecting them. Such oospores are also subcentric. The reserve substance of an eccentric oospore is a single refractive globule situated laterally to the cytoplasm, with the ooplast contiguous to it on one side, and the plasmalemma bounding it on the opposite side. Thus what has been universally looked upon as a "cap"of cytoplasm over the single large reserve globule in an eccentric oospore is in reality the ooplast. Dick introduced a new terminology to characterize an eccentric oospore, referring to its parts as phases or fractions. The ooplast is evidently what he refers to as an optically inactive clear matrix.

It has been suggested that the eccentric nature of an oospore is favored if the zygote matures slowly. Drechsler was prompted to include fine particles of maize in plated media since these bits of solid substratum appeared to him to enhance development of the oogonia of Aphanomyces cladogamus thus insuring that the oospores would have "the correct internal organization necessary for longevity". The implications in Newby's and Drechsler's comments warrant exploration.

Howard and Moore described a fourth type of oospore, the expersate one. Such a zygote possessed a single reserve globule as did the eccentric one, but in contrast to the latter, the refractive material was contained entirely within the matrix of the cytoplasm (not just bounded by the plasmalemma as in the case of the eccentric type), and there was no ooplast. Species of Aphanomyces were alleged to have expersate oospores. Howard reported that the surface of the refractive globule in oospores of A. laevis was irregular as if membrane-bound, and might not be lipid. At one point in the germination of an eccentric oospore, however, the refractive reserve body becomes convoluted as it changes chemically.

On the basis of brightfield observations, Dick concluded that each oospore of Aphanomyces stellatus and A. cladogamus in fact has an ooplast although this structure is optically inactive. The absence of supportive illustrations notwithstanding, we agree with Dick that a configuration interpreted as expersate cannot be recognized as a separate entity. It must be admitted, however, that the description by V. D. Matthews of the oospores in Leptolegnia eccentrica surely recalls the "expersate type".

In 1960(c) Dick introduced the term subeccentric for oospores of a watermold which he at that time identified as Saprolegnia asterophora, but later placed in Scoliolegnia†. In a subeccentric oospore (the position of the ooplast with respect to the other fractions in the cell has not been determined) the reserve deposit consists of several globules of various diameters clustered together at one side of the cytoplasm or, in some cases, on opposing sides. Such oospores have been detected in Pythiopsis cymosa, Calyptralegnia achlyoides, C. ripariensis, and Isoachlya eccentrica†. It must be recorded that during maturation some eccentric oospores may show temporarily a subeccentric configuration as the reserve globules coalesce. Hence, the concept of subeccentric as a valid oospore type rests solely on the assumption that this condition represents a fully mature zygote.

The subcentric Type III oospore configuration described and illustrated by T. W. Johnson for two isolates of Achlya treleaseana has not again been reported. It clearly is an aberrant form undeserving of continued recognition. The oospore structure described for Leptolegnia subterranea by Coker and Harvey as "a cup of oil globules on one side" is best considered as subeccentric.

On the basis of oogonium and oospore morphogenesis Dick proposed that the Oomycetes can be distributed into three groups. The Saprolegniaceae constitute a distinct group characterized by centrifugal oosporogenesis without the formation of periplasm, and aplerotic oospores containing a peripherally disposed reserve. Species of

Leptolegnia perhaps constitute a cluster of taxa to be excluded from the Saprolegniaceae, Dick suggests, because in these, oosporogenesis is said to be centripetal.

The reserve globules, by whatever name one chooses to apply, have figured inextricably in taxonomy of the watermolds. Traditionally, the oospore type has been held to be inviolate within a given species. This is not an entirely acceptable conclusion since in some taxa - Saprolegnia ferax and Achlya racemosa, to name two - centric and subcentric oospores may occur within the same oogonium. Instances of eccentric oospores and centric or subcentric ones in the same oogonium (or species) have not been reported.

Oospore Germination

With some very scattered exceptions, instances of oospore germination in members of the family have been noted largely by chance. It has long been thought that zygote germination was accompanied by meiosis, but current evidence points to the contrary. Two main aspects of germination have been explored: The factors triggering or influencing the events, and the morphological products of the process.

In 1899, Trow recorded some details of oospore germination in Achlya americana var. cambrica†, and although later studies by others contradict his account in a few minor ways, his treatment still describes reasonably correctly morphogenesis of germination. At least three events appear to take place as germination commences in an eccentric oospore. The single, large, reserve globule first becomes embedded in the matrix of the cytoplasm (the complex deposit of lipids is already within the plasmalemma, but exterior to the ooplast, so that the refractive material appears to migrate into the center of the oospore). Once within the matrix, the globule begins to diminish in diameter, and as it does so, its outline becomes irregular. Trow simply said that the globule was absorbed. Concurrently, the oospore wall becomes progressively thinner, and the zygote itself increases in diameter. Both Trow and Weston reported that during these events the protoplasm actually increased in volume, but precisely how this occurs and what internal changes make it possible have not been determined. In any case, prior to emergence of the germ hypha (more than one may develop in some species) the cytoplasm becomes centrally vacuolate. As the young hypha elongates, its vacuole becomes confluent with that in the oospore. With continued growth, the hypha eventually breaches the confining oogonial wall (if that structure is still intact), but the details of this process are obscure. It has been firmly established that mitotic divisions of the somatic nuclei accompany germ tube growth.

So far as we are aware, the path of germ tube egress from the oogonium described by F. R. Jones and Drechsler for Aphanomyces euteiches has not again been observed. In the germination of some oospores in this species the germ tubes breached the fertilization tubes and grew through them, thence into the antheridial cell. Subsequently, these young hyphae ruptured this cell and became extramatrical. The fertilization tube in these cases functioned as a channel for the emerging germ hypha.

At what point in its existence is the oospore capable of germination? Clausz demonstrated that oospores of Achlya hypogyna matured in eight days, but there followed a lengthy afterripening period at the end of which the irreversible events in germination commenced. This "ripening" sequence is not necessarily universal in the watermolds if Latham's observations are correct; he reported that the oospores of A. recurva, were capable of germination immediately after they were formed (in about three days' time). In one of the first recorded comments on oospore germination, Pringsheim remarked that there was no fundamental difference between a fertilized oospore and one matured parthenogenetically, except that the latter germinated more readily than the former.

The one study of oospore germination in watermolds that is still accepted as the most extensive of its kind was that by A. W. Ziegler. He recognized four patterns of germination among the several species he examined, namely, the production of (1) A single germ hypha (of variable length) bearing an apical sporangium. (2) A sparingly branched mycelial system forming sporangia at the tips of the main branches. (3) A branched mycelium lacking sporangia. (4) A single, unbranched hypha. In addition to these modes of germination a fifth may be added (though it is evidently quite uncommon and may be abnormal). De Bary, and F. R. Jones and Drechsler among others described oospores which simply converted directly into sporangia without an intervening hyphal system. Trow, on the contrary, had concluded that such a pattern of germination never occurred.

Little is known about the segregation of sexual strains in the germination of zygotes in dioecious species, but experimental work by J. N. Couch with germ hyphae from oospores of Dictyuchus monosporus is pertinent. He succeeded in germinating the oospores of this species, cut off segments of the resulting mycelium, and propagated colonies from these. The results of his experiments, in terms of mating strains produced, demonstrated that the developing mycelium apparently consisted of certain sexual or nonsexual regions. Some parts of the mycelium were pure male, some pure female, and others were undifferentiated sexually. As growth proceeded, the "sexes" became intermixed.

Precisely what products oospores form in nature when they germinate, and how those products are generated is not known. Presumably motile spores produced by germinating oospores provide the inoculum for substrates when dried soil is flooded and baited. This being so, all functional oospores -including those with the nonsporangiate germination patterns recognized by A. W. Ziegler - must at some point be able to develop planonts. The types of zygote germination as they are now known are products of the laboratory, and the process may bear little resemblance to reality.

Some work on Aphanomyces euteiches by F. R. Jones and Drechsler, Scharen and Olofsson, and on A. raphani by Ghafoor bears directly on "natural" germination of oospores. By embedding oospores in plant debris in cellophane bags, and positioning these containers near pea roots, Scharen found that oospore germination was stimulated. Pea root extracts, horse dung infusion, and pea leachate in sand also provided a stimulatory effect. Oospores did not sprout in distilled water, and unsterilized soil had a distinctly fungistatic effect (there were significantly higher germination percentages in sterile than in nonsterile soil). Using oospores of A. euteiches harvested from naturally infected pea roots Olofsson demonstrated that pH was a significant factor in germination in these species. Oospores in a medium at or about neutrality (6.5-7.0) were not stimulated to germinate by heat (60 °C), subjecting them to below freezing temperatures (-4.0 and −20 °C), or alternately freezing and thawing them. Similarly, there was no germination of oospores submerged in tap water adjusted to a pH below 3.2 or above 5.1. Germination took place, however, in water in the range of pH 4.5-4.8. Oospores produced germ hyphae more readily if they had been harvested from decayed roots (tissue about four months old) than if taken from undecayed roots (about four weeks old), Olofsson suggested that at a low pH fungistatic substances in pea roots were inactivated, and oospore germination was then no longer inhibited. When F. R. Jones and Drechslerplaced oogonium-bearing hyphae from 15-day-old cultures (A. eutieches) in Van Tieghem cells oospore germination occurred. In these cases, therefore, pH evidently was not a contributing factor. Ghafoor found that the germination percentage of oospores of A. raphani was more than doubled when radish seedlings were grown in autoclaved soil containing these oospores than when the host was not present in the sterile soil.

It is evident that much variation exists among the species of watermolds with regard to the germinability of their oospores under laboratory conditions. Moreover, the time required for the process to begin once the "aged" oospore is capable of germinating is vastly different among the representatives of the family.

Autospore

An autospore is a non-motile (non-flagellated) spore that is produced within a parent cell, and has the same shape as the parent cell, before release. Autospores, in addition to zoospore and aplanospore, are one of the three types of spores that algae use to reproduce and spread asexually. Autospores occur in several groups of algae, including Eustigmatophyceae, Dinoflagellates and green algae. For example, the colonial alga Dichotomococcus produces two autospores per reproducing cell; the autospores escape through a slit in the cell wall and remain attached to the mother cell.

Autospore Formation

An autospore is defined one of the daughter cells formed by the internal division of a single cell. Autospores are formed as a result of fission in the mitotic phase of cell division of green algae. There are multiple methods that a cell can take to form autospores. The cell can undergo a multiple fission after 2 nuclear divisions where 4 autospores will form which is the preferred mechanism in organisms such as P. subcapitata. In addition to this mechanism, there is binary fission (the split into 2 autospores) also known as the "two-autospore type" and the "eight-autospore type".

Ballistospore

A ballistospore or ballistoconida is a spore that is discharged into the air from a living being, usually a species of fungus. With fungi, most types of basidiospores formed on basidia are discharged into the air from the tips of sterigmata. At least 30 thousand species of mushrooms, basidiomycete yeasts, and other fungal groups may discharge ballistospores, sometimes at initial accelerations exceeding 10 thousand times *g*.

Zoospore

Several groups of fungi and fungus-like organisms produce motile, flagellate spores. These zoospores differ in number and types of flagella and in ultrastructural details in the different groups of organisms. But all zoospores share some common features:

- They are naked, wall-less cells, specialized for dispersal because they cannot divide nor absorb organic nutrients.
- They swim for many hours, using endogenous food reserves, then encyst by retracting or shedding their flagella and secreting a wall.
- They can respond to a range of environmental signals, which they use to locate the sites where they will encyst.

This precise site-selection or "homing" behaviour accounts for much of the interest in zoosporic fungi. In practical terms, it gives zoospores an important role in infection of plant and animal hosts, and sometimes as vectors of major plant viruses.

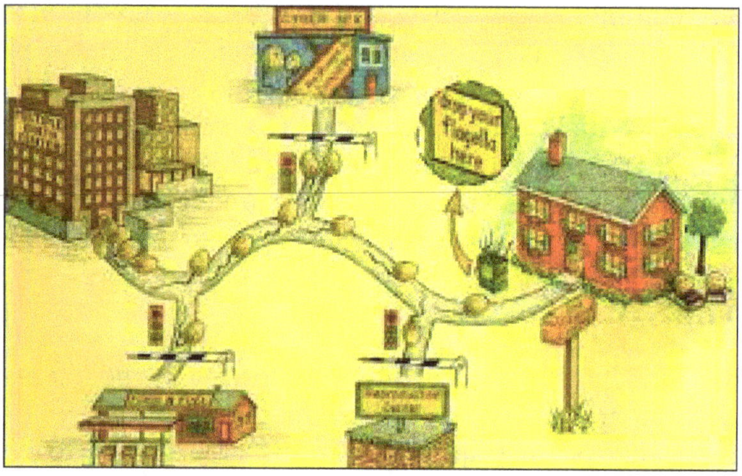

Homing sequence of zoospores of the oomycota, e.g. Pythium or Phytophthora species, Zoospores released from the 'Multiple Birth Center' are barred access to Food and Fuel, Cyber Sex (no minors or zoospores admitted) and the Reproduction Center. At the end of their quest, they drop their flagella into the bin and encyst at their new home.

Table: Some economically significant zoosporic fungi and related organisms.

Plasmodiophorids: Produce small zoospores with two anterior whiplash flagella.	
Plasmodiophora brassicae	Major pathogen of cruciferous crops, causing clubroot disease.
Polymyxa graminis, P. betae	Zoosporic vectors of some important plant viruses, including beet necrotic yellow vein virus.
Spongospora subterranea	Causes powdery scab of potato and is a vector of potato mop-top virus.
Chytridiomycota: produce small zoospores with typically a single posterior whiplash flagellum.	
Olpidium brassicae, O. radicale	Vectors of several plant viruses.
Synchytrium endobioticum	Causes potato wart disease.
Catenaria anguillulae	Parasite of nematodes and fluke eggs.
Coelomomyces spp.	Parasites of aquatic larvae, including mosquitoes.
Neocallimastix spp.	Obligately anaerobic saprotrophs in the rumen, active in cellulose breakdown.
Oomycota: Produce large, kidney-shaped zoospores with an anterior tinsel flagellum and a posterior whiplash flagellum.	
Saprolegnia and Achlya spp.	Freshwater saprotrophs, but some are parasites of salmonid fish in fish farms.
Aphanomyces spp.	A. astaci is a serious pathogen of crayfish, A. euteiches of pea crops, A. cochlioides of spinach crops, etc.
Pythium spp.	Major seedling pathogens of crops; also cause decline diseases of orchard trees.
Phytophthora spp.	Major pathogens of crops, including P. infestans (potato blight), P. parasitica (glasshouse crops, etc.), P. cinnamomi (avocado, pines and many woody hosts).
Downy mildew fungi	Several major pathogens, including Plasmopara viticola on grapevine.

Zoospores of the Oomycota

Structure and Ultrastructure

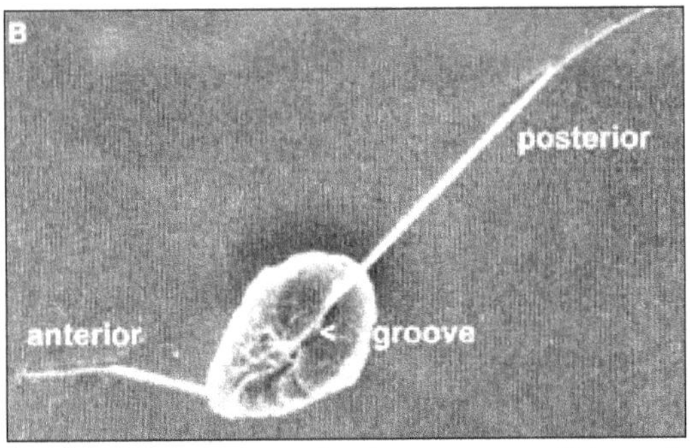

Scanning electron micrograph of a zoospore of Phytophthora.

The zoospore body (soma) of the oomycota is typically kidney-shaped, with a ventral groove. Two flagella emerge about one-third of the distance along this groove. The posteriorly directed flagellum is long and smooth - a "whiplash-type" flagellum. The anterior flagellum is a shorter "tinsel-type" which bears tripartite hairs (mastigonemes) along its length.

When viewed by transmission electron microscopy these zoospores show a distinct zonation of organelles. Close to the ventral groove is the nucleus, which extends as a beak towards the point where the flagella are inserted. In a different plane of focus (not shown) this region also contains a conspicuous water-expulsion vacuole - an osmoregulatory apparatus. The zoospore also contains many mitochondria, and vacuoles with striated contents - the "fingerprint vacuoles". Just beneath the cell surface are three types of peripheral vesicle - the large peripheral vesicles, small dorsal vesicles, and small ventral vesicles. The zoospore is surrounded by a cell membrane, beneath which is a series of plate-like membranes - the peripheral cisternae.

Only a small part of each flagellum is seen in figure below. The flagella have a central core termed the axoneme, composed of microtubules and associated proteins. Over most of its length, the axoneme is surrounded by a membrane which is continuous with the cell membrane. The mastigonemes of the tinsel flagellum are glycoprotein extensions which project through the flagellar membrane.

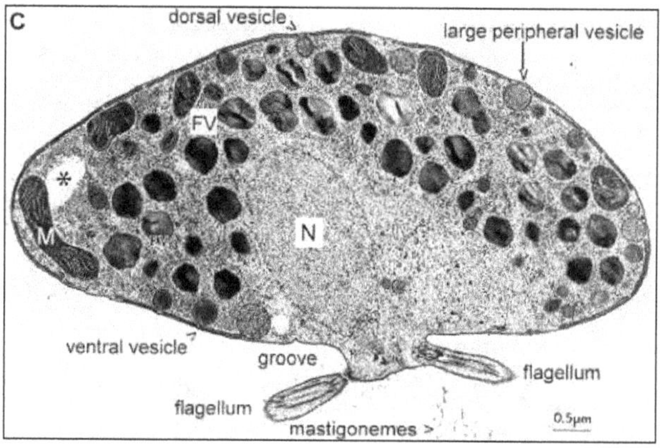

Transmission electron micrograph of a Phytophthora zoospore (about 15 micrometres diameter) prepared by freeze-substitution and sectioned through the region where the nucleus (N) extends towards the flagellar basal apparatus. The spores contain several mitochondria (M) and fingerprint vacuoles (FV). Three types of peripheral vesicle are shown, and flattened peripheral cisternae lie beneath most of the zoospore plasma membrane. The regions marked with an asterisk are thought to be where lipids were removed by solvents during fixation.

Zoospore Swimming

Zoospores can swim for many hours, at 150 micrometres or more per second, so in theory they could disperse several metres from their point of release by their own activities. But in practice this seldom happens because they change direction frequently when they meet obstacles or by spontaneous, random turns. They are best suited for local dispersal but can be carried longer distances in moving water. In addition, some plant-pathogenic species (e.g. Phytophthora infestans, which causes potato blight) have wind-dispersed sporangia and these release zoospores when they land on a host surface.

The swimming mechanism is only partly understood. The flagellate bacteria have a rotor which causes the flagella to rotate, and spontaneous turns are achieved by a reversal of the rotor direction. In zoospores there is no flagellar rotor. Instead, the flagella beat in a sine wave which moves from the base to tip of each flagellum, and the zoospore swims in a straight helix, rotating about its axis. Hydrodynamic studies suggest that the posterior flagellum serves mainly as a rudder, while the anterior flagellum contributes up to 90% of the swimming speed because the mastigonemes act like oars, converting the forwards-directed sine wave to a backwards thrust.

The corkscrew-like pattern of zoospore swimming, interspersed with random changes of direction, is shown in figure below where the tracks of Pythium zoospores were photographed with a 5-second exposure, using darkfield microscopy. However, in this case many of the abrupt turns were caused when zoospores hit the surface of a glass coverslip.

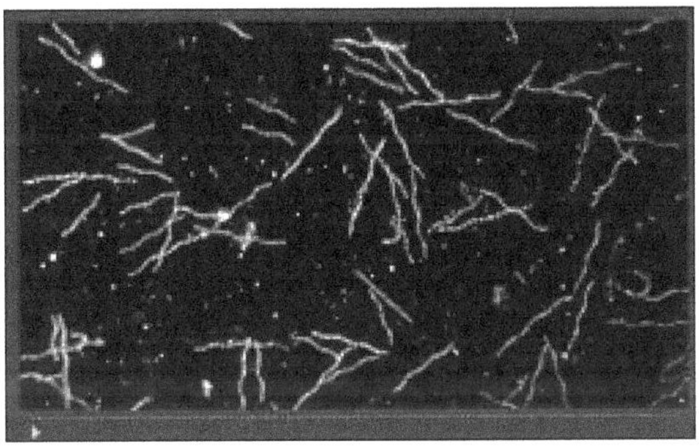

The regulation of swimming pattern is strongly influenced by calcium, which regulates the behaviour of flagella or cilia in other eukaryotic cells. For example, when external calcium is removed by a chelator (EGTA) then the zoospores swim in almost straight lines. They swim in circles when treated with substances (e.g. lanthanum or verapamil) that block calcium channels in the cell membrane. They swim in a skidding fashion when treated with compounds (e.g. trifluoperazine or dibucaine) that interfere with the calcium-binding protein calmodulin. And they swim in an erratic, jerky manner when treated with an ionophore (which equilibrates the calcium concentration across the cell membrane) or with amiloride (which disrupts ion exchange across the membrane).

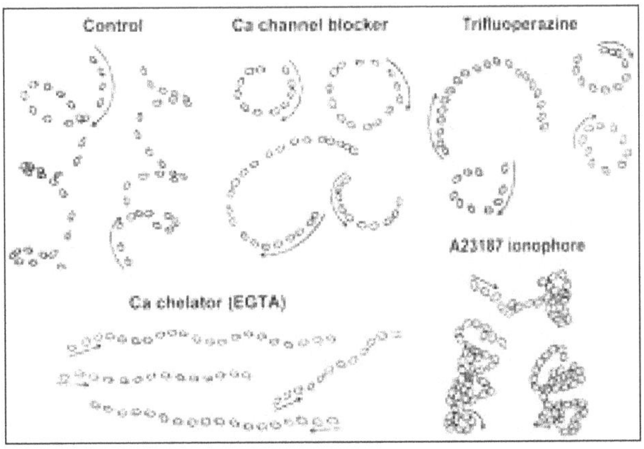

Zoospore tracks of Pythium aphanidermatum treated with various calcium-interfering substances (Control = no treatment). Each track shows the position of a single zoospore at intervals of 0.1 sec (control, channel blocker), 0.2 sec (EGTA, ionophore) or 0.3 sec (trifluoperazine) traced from a video screen; arrows show the initial direction of movement.

It is still unclear exactly how these substances exert their effects on motility, but two general points are of interest. First, zoospores treated with any of these substances do not show the normal, periodic changes of direction and also cannot respond to attractants. So the normal swimming pattern is essential for the role of zoospores in site-selection, discussed below. Second, these findings illustrate one of the many unifying themes of biology, because calcium plays a central role in all eukaryotic organisms. In fact, some of these substances that affect zoospores by interfering with calcium-mediated processes are used as pharmacological agents to regulate heartbeat, kidney function or psychological disturbances. A nice demonstration of this was reported recently by NASA scientists, who tested the effects of various substances on the web-building activities of spiders. This serious piece of research showed that caffeine was one of the most disruptive agents. Caffeine exerts its effects by depleting the calcium levels in intracellular stores.

Caffeine strongly disrupts normal zoospore function, and ethanol is a strong chemoattractant of some zoospores.

The Zoospore Homing Sequence

Having laid the foundations of zoospore structure and function, we now consider the homing sequence leading to infection of a host from a zoospore. The following diagram gives a composite view of this sequence, starting from a sporangium, such as one of the wind-dispersed sporangia of Phytophthora infestans (the cause of potato blight). However, it should be noted that this composite view has been built up from studies on various Phytophthora, Pythium and Aphanomyces species. Not all of these stages may apply to any one fungus.

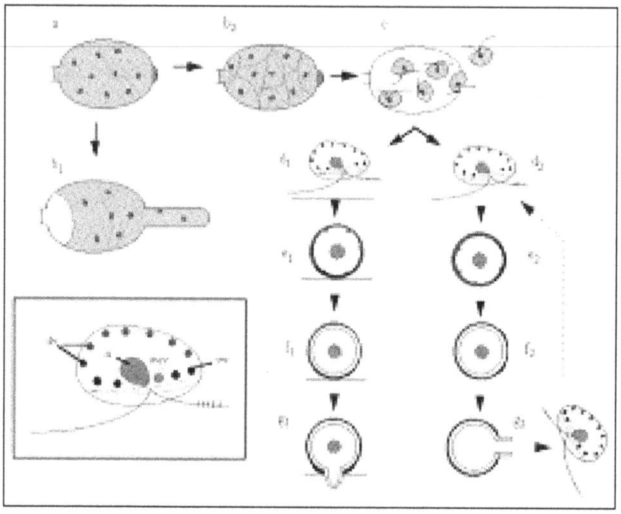

Stage a (top left) in this diagram shows a sporangium of a Phytophthora infestans. In this fungus the sporangium can germinate in two ways: by hyphal outgrowth (b_1) or by undergoing protoplasmic cleavage (b_2) to release zoospores (c) by dissolution of the sporangial papillum. Zoospores (inset at bottom left) have an anterior tinsel flagellum and a posterior whiplash flagellum, inserted in a ventral groove close to the position of the nucleus (n) and water-expulsion vacuole (wev). Small dorsal vesicles (dv) and small ventral vesicles (vv) lie just beneath the plasma membrane. The sequence d1 to g_1 shows the events when a zoospore is induced to encyst by a host surface component. The spore typically orientates, settles and adheres with the future, pre-determined germination site against the host. (e_1). Cyst coat glycoprotein is then released by exocytosis of the small dorsal vesicles, and proteinaceous adhesive is released from the ventral vesicles. Over the next few minutes a microfibrillar cyst wall is synthesised beneath the cyst coat (f_1). Then after 20-30 min the cyst germinates by a hyphal outgrowth towards the host (g_1); the hypha either infects directly or (not shown) produces a swollen appressorium before infection. An alternative sequence (d_2-g_2) occurs if a zoospores does not find a host. It encysts eventually and, after some hours, typically releases a further zoospore.

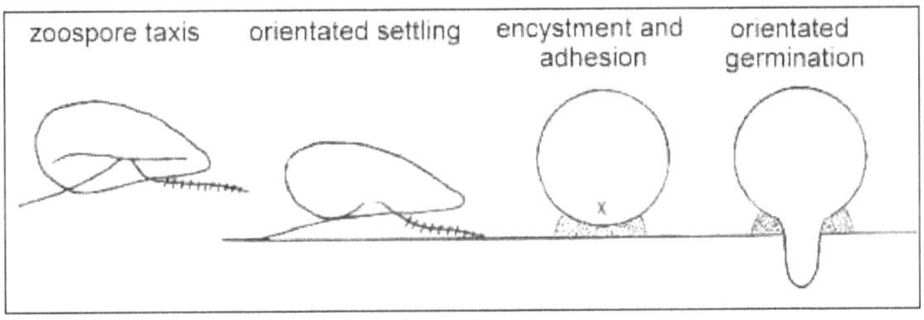

- Accumulation of zoospores at specific sites by sensing gradients of attractants or repellents (zoospore taxis).
- Settling and orientation on the host surface, perhaps by recognition of specific host surface components.
- Adhesion and encystment, involving the release of adhesins and production of a cyst wall.
- Germination with a fixed orientation, from an apparently predetermined point.

These four stages are remarkable for many reasons:

- They occur so rapidly that they can lead to infection from an originally motile spore within 30-40 minutes.
- During his short time the cells undergo major changes of organisation - the originally wall-less, motile cell is transformed into a walled cyst which then gives rise to a new hyphal apex.
- The cells can respond to many exogenous cues, including those that change the direction or speed of swimming, others that trigger encystment and yet others that can trigger cyst germination.

- Throughout this time the cells do not take up external organic nutrients - this occurs only after the cyst has germinated and the germ-tube has grown to some length.

Thus, we can think of zoospores as being largely pre-programmed cells, which require only the appropriate signals to progress through a complex sequence of events leading to infection.

Zoospore Taxis

The term taxis can be used generally to describe a change in swimming orientation of a motile cell in response to an external stimulus. The zoospores of various fungi have been shown to respond to chemical gradients (chemotaxis), oxygen (aerotaxis), electrical or ionic fields (galvanotaxis, electrotaxis) or light (phototaxis). Some also can swim against a water current (rheotaxis) and some swim upwards in suspension (negative geotaxis) but probably not by sensing gravity as such. In addition to all these factors, when zoospore suspensions are sufficiently dense then the cells can autoaggregate - a phenomenon termed adelphotaxis and perhaps caused by the release of an attractant substance.

One or more of these factors can cause zoospores to accumulate at specific sites. For example, the zoospores of plant pathogenic fungi commonly accumulate at wound sites, or near the stomata of leaves, or near the root tips. Similarly, the zoosporic parasites of small animals often accumulate at the body orifices.

Zoospore taxis can also serve to bring zoospores to sites where they will be destroyed by toxins. A natural example of this is seen in the responses of zoospores to oat roots: see zoospores and saponins.

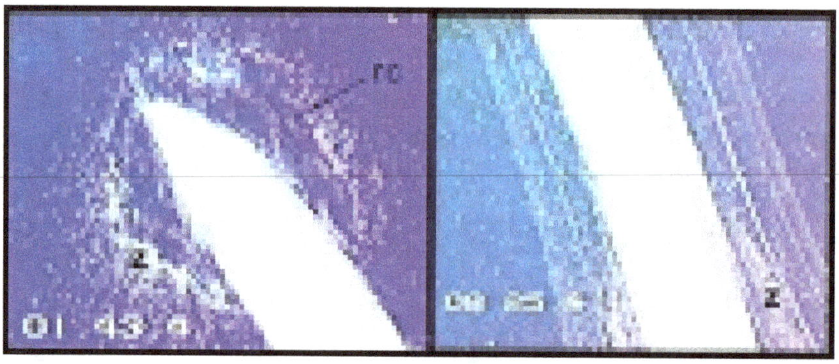

Pythium zoospores accumulating on wheat roots: at 1 min 43 sec (left) and 8 min 25 sec (right) after roots were immersed in zoospore suspension. The left image shows a root tip surrounded by a ball of root tip mucilage containing shed root cap cells (rc). A mass of zoospores (z) has accumulated over the surface of the root tip mucilage. The right image shows a region just behind the tip of a root coated with two layers of alginate gel. Zoospores (z) have accumulated locally on the alginate in the zone where root hairs start to emerge.

Capillary model systems can be used to investigate the individual attractants of zoospores and thus give clues to the factors involved in host-location. Typically, the zoospores of root-infecting Pythium and Phytophthora species are attracted strongly to a few individual amino acids (e.g. aspartic or glutamic acid), sugars (e.g. glucose) or volatile compounds (e.g. ethanol, aldehydes), all of which are common components of root exudates. But the spores respond better to mixtures than to individual compounds. In addition, a few examples of host-specific attractants have been reported from capillary studies. For example, Phytophthora sojae, a host-specialised pathogen of soybean roots, is strongly attracted to the soybean flavonoids daidzein and genistein, and Aphanomyces cochlioides (from spinach) is strongly attracted to the host flavonoid, cochliophilin A. These reports are interesting because they parallel the behaviour of host-specific Rhizobium species, which also are attracted to the flavonoids of their hosts and can be repelled by non-host flavonoids.

It is assumed that the responses to different attractants are mediated by chemoreceptors in the zoospore or flagellar membrane. But no receptor has yet been purified and characterised from a zoosporic fungus.

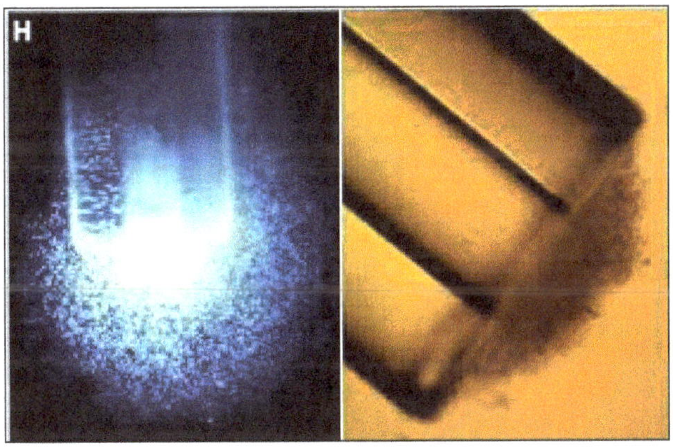

Left: zoospores of Phytophthora palmivora accumulating at the mouth of a capillary filled with a fluorescently tagged amino acid (dansyl-asparagine) Right: zoospores of Pythium aphanidermatum accumulated and encysted at the mouth of a capillary containing malt-extract agar. Germ-tubes have grown from the cysts into the agar in the capillary.

Zoospore Encystment

Zoospores can be induced to encyst by artificial agents such as high concentrations of chemicals or mechanical agitation in laboratory conditions. But in nature they probably respond to recognition of a host surface component. For example, the zoospores of many Pythium and Phytophthora species are induced to encyst by pectin or by root surface mucilage (which often has a similar composition to pectin). There may also be a degree of specific recognition. For example three species of Pythium responded differently to a range of plant-derived polymers in laboratory assays, and the Pythium species that characteristically infect the roots of grasses and cereals show only a weak ability to encyst on non-grass roots.

Table: Percentage of zoospores induced to encyst by addition of different polymers; significant differences from the controls (no treatment) are shown in bold. Some treatments are shown only as "yes" (encystment induced) or "no" (no effect).

Polymer	Fungus		
	P.aphanidermatum	P.catenulatum	P.dissotocum
Control (none)	10	12	25
Arabinoxylan	61	18	26
Methylglucuronoxylan	10	54	23
Xyloglucan	10	17	73
Fucoidan	7	54	25
Mixed linkage glucan	11	45	39
Gum arabic	37	69	65
Alginate	Yes	Yes	Yes
Cellulose	No	Yes	Yes
Chitin (crab shell)	Yes	No	Yes

Table: Numbers of zoospores that encysted on seedling roots of wild grasses or wild dicotyledonous plants in replicated laboratory tests.

	Pythium graminicola			Pythium aphanidermatum		
	Grass (A)	Dicot (B)	A/B	Grass (A)	Dicot (B)	A/B
Expt 1	279	28	10.0	248	243	1.1
Expt 2	112	44	2.6	94	103	0.9
Mean of 9 experiments			3.6			1.0

Seedlings of wild grasses and dicotyledonous plants were collected from field sites and immersed in zoospore suspensions of either P. aphanidermatum (which has a characteristically broad host range) or P. graminicola (which characteristically infects Gramineae). In each of 9 experiments with replications within the experiments, similar numbers of P.aphanidermatum zoospores encysted on grass and dicot roots, whereas P. graminicola encysted mainly on grass roots. Presumably P. graminicola zoospores recognise components of the root surface mucilage of Gramineae, which is substantially different from the root mucilage of other plants.

Several events occur in quick succession during zoospore encystment, suggesting that the cells are pre-programmed to respond to an initial encystment signal.

- The flagella are shed, the zoospore rounds off, and the nucleus migrates to a central position.

- Some of the peripheral vesicles fuse with the cell membrane to release their contents onto the cell surface. These contents include a cyst coat glycoprotein from the small dorsal vesicles and an adhesive protein or glycoprotein from the small ventral vesicles.
- The flat peripheral cisternae start to bud vesicles which will deposit a cyst wall underneath the cyst coat.
- The production of the cyst wall is usually completed in about 5-6 minutes. Then the water-expulsion vacuole disappears, and a germ-tube emerges about 20-30 minutes later, to produce either a hypha or an appressorium (pre-infection swelling).

The disruption of any of these events could potentially lead to successful control of disease caused by zoosporic fungi.

Zoospore Alignment and Orientation of Germination

When zoospores settle on a host surface, the cysts almost invariably germinate towards the host. This raises the possibility that zoospores might pre-align during encystment. Several lines of evidence support this idea and also suggest that the site of germination is predetermined. In other words, a zoospore settles and encysts on a host with a precise orientation (docking) so that the emerging hypha can rapidly penetrate the host.

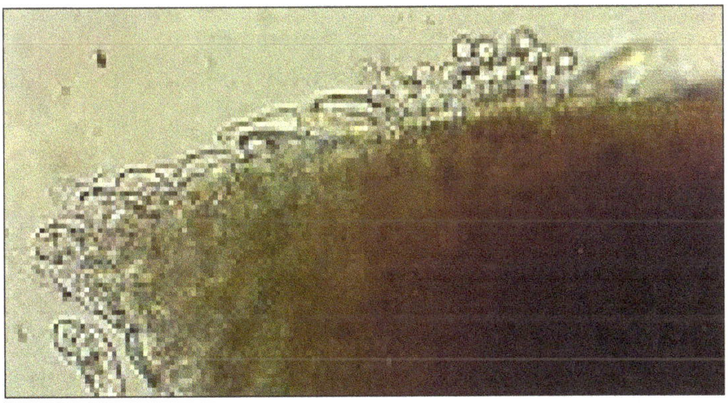

Zoospore cysts of Pythium aphanidermatum germinating towards a wheat root after the zoospores settled and encysted just behind the root tip.

The first evidence for pre-alignment came from simple laboratory experiments with Pythium zoospores, shown diagrammatically in figure. Motile spores were placed in a chamber with a block of nutrient-rich agar (malt-peptone agar) at one end and a block of nutrient-free water agar at the other end. Most of the zoospores swam towards the nutrient-rich agar, then encysted and germinated towards the agar block. Fewer zoospores settled and encysted near the water agar, and these cysts germinated with random orientation.

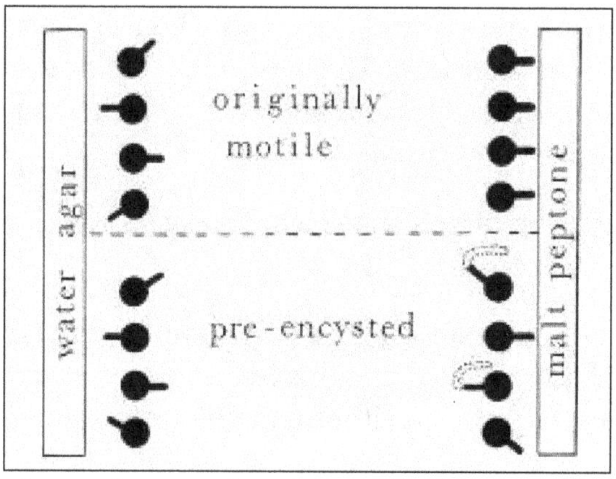

Simple experiment to show that zoospores of Pythium have a fixed (predetermined) point of germination.

These findings could be explained in two ways:

- Either the zoospore has a fixed (predetermined) site of germination which is located next to the attractant source (e.g. host) by pre-alignment of the zoospore when it encysts.
- Or the zoospore cyst has the potential to germinate from any point, and the actual site of germination is influenced by the attractant source (e.g. host).

To distinguish between these possibilities, zoospores were encysted artificially (by agitation) then placed in a chamber. The pre-encysted cells always germinated with random orientation, even when they were positioned next to the nutrient-rich agar. So they seem to have a fixed site of germination which cannot be influenced by external factors such as nutrients. Nevertheless, the germ-tubes can change direction after they have emerged from a fixed position, and then grow towards a nutrient source.

Further understanding of these events has come from the use monoclonal antibodies (MAbs) raised against components of zoospores of Phytophthora cinnamomi. Some of these MAbs are specific to the glycoprotein contents of dorsal vesicles in the zoospores. By MAb-tagging, this glycoprotein was shown to be released by exocytosis during encystment and it forms the cyst coat. Other MAbs are specific to the protein or glycoprotein contents of ventral vesicles, which line the shoulders of the zoospore ventral groove. This material also was released during encystment, and was shown to be deposited next to a host surface, where it is thought to function as an adhesive. Thus, the zoospores must pre-align during encystment, with the ventral groove next to a host surface.

This raises the possibility that the flagella might be involved in recognising host surface components, and again evidence has come from the use of MAbs. Of all the antibodies raised against zoospores of P. cinnamomi, only one caused rapid encystment when added to motile spores. This MAb binds specifically to the surface of both flagella, and it also induced rapid encystment of other Phytophthora and Pythium species. In contrast,

MAbs that bind to the whole zoospore surface or to components of only the anterior flagellum did not cause rapid encystment.

Aplanospore

Aplanospores are non-motile asexual spores found in certain algae and fungi. They are formed from a part of a protoplast of a vegetative cell. Aplanospores consist of a cell wall, which is different from the original parental cell wall, allowing the survival of the organism in harsh conditions. Haematococcus Pluvialis forms cell aggregates due to the nitrogen deprivation. These aggregates finally become aplanospores.

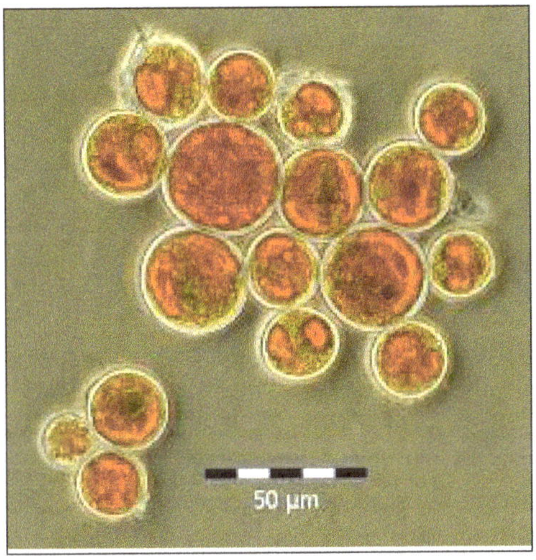

Aplanospores of Haematococcus Pluvialis.

Difference between Zoospores and Aplanospores

- Definition:
  - Zoospores: A zoospore is a motile asexual spore produced by certain algae, fungi, and protozoans.
  - Aplanospores: An aplanospores is a nonmotile asexual spore produced by certain algae and fungi.
- Occurrence:
  - Zoospores: Zoospores occur in Phycomycetes.
  - Aplanospores: Aplanospores occur in green algae.
- Motility:
  - Zoospores: Zoospores are motile.
  - Aplanospores: Aplanospores are nonmotile.

- Cell Wall:
    - Zoospores: Zoospores do not possess a true cell wall.
- Size:
    - Zoospores: Zoospores are small in size.
    - Aplanospores: Aplanospores are large compared to zoospores.
- Significance:
    - Zoospores: Zoospores are incapable of undergoing harsh environmental conditions.
    - Aplanospores: Aplanospores are capable of undergoing harsh environmental conditions due to the presence of a cell wall.
- Examples:
    - Zoospores: Oomycota like Phytophthora, Chytridiomycota, Myxomycota and Plasmodiophoromycota as well as Opisthokonts and heterokonts produce zoospores.
    - Aplanospores: Lower fungi, algae like Haematococcus Pluvialis, Chlamydomonas and Vaucheria produce aplanospores.

References

- Reproduction-in-fungi-vegetative-asexual-and-sexual-methods-7296: yourarticlelibrary.com, Retrieved 25 February, 2020
- Fungal-spores: mytrainingcourses.ca, Retrieved 23 July, 2020
- Hawksworth DL. (2013). Ascomycete Systematics: Problems and Perspectives in the Nineties. Springer. p. 116. ISBN 978-1-4757-9290-4
- Germination-of-zygospore-in-lower-fungi-botany, fungi-63123: biologydiscussion.com, Retrieved 15 April, 2020
- Pringle, A.; Patek, S. N.; Fischer, M.; Stolze, J.; Money, N. P. (1 July 2005). "The captured launch of a ballistospore". Mycologia. 97 (4): 866–871. doi:10.3852/mycologia.97.4.866. PMID 16457355
- Difference-Between-Zoospores-and-Aplanospores-317011308: researchgate.net, Retrieved 18 June, 2020
- Hsueh, Y. P.; Heitman, J. (2008). "Orchestration of sexual reproduction and virulence by the fungal mating-type locus". Current Opinion in Microbiology. 11 (6): 517–524. doi:10.1016/j.mib.2008.09.014. PMC 2622727. PMID 18935978

Fungal Diseases

CHAPTER 6

Fungal diseases occur when a harmful fungus enters the immune system of the body. Dermatophytosis, Fungal keratitis, sporotrichosis, blastomycosis, histoplasmosis, aspergillosis, coccidioidomycosis, paracoccidioidomycosis, etc. are a few of the fungal diseases. This chapter discusses the symptoms, sources, diagnosis and treatment of various fungal diseases in detail.

MYCOSIS

Mycosis is a disease caused by any fungus that invades the tissues, causing superficial, subcutaneous, or systemic disease. Superficial fungal infections, also called dermatophytosis, are confined to the skin and are caused by Microsporum, Trichophyton, or Epidermophyton; athlete's foot, for example, is caused by Trichophyton or Epidermophyton. Subcutaneous infections, which extend into tissues and sometimes into adjacent structures such as bone and organs, are rare and often chronic. Candidiasis (Candida) may be a superficial infection (thrush, vaginitis) or a disseminated infection affecting certain target organs, such as the eyes or kidneys. Painful ulcerations and nodules appear in subcutaneous tissues in sporotrichosis (Sporothrix). In systemic fungal infections fungi may invade normal hosts or immunosuppressed hosts (opportunistic infections). Cryptococcosis (Cryptococcus) and histoplasmosis (Histoplasma) are marked by respiratory distress.

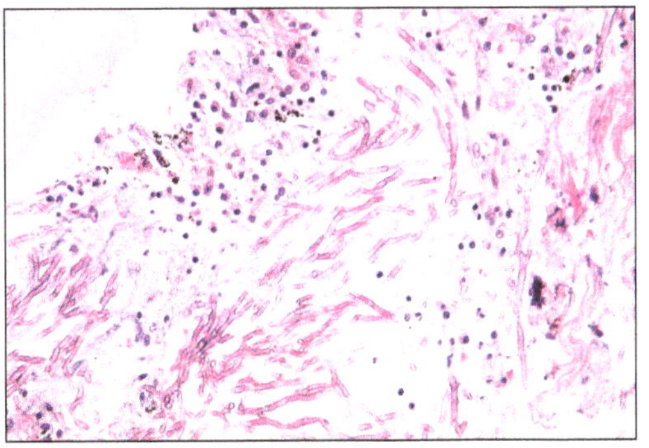

Micrograph showing hyphae of the fungus Aspergillus in cells of the pulmonary system. Pulmonary aspergillosis is a type of mycosis.

Effective therapy against the invasive fungi is limited because the same antibiotics that interfere with fungi also attack the host's cells. Griseofulvin has met with some success in the treatment of superficial mycoses. Amphotericin B and flucytosine have also been used in treating the subcutaneous and systemic mycoses.

The following clinical groupings of Mycosis may be recognized:

Superficial Mycoses

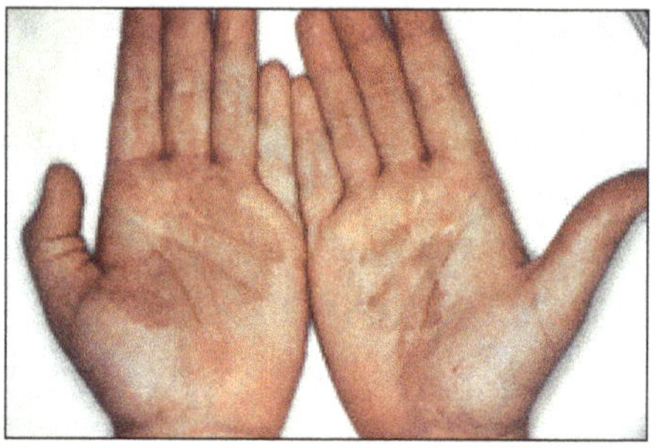

These are superficial cosmetic fungal infections of the skin or hair shaft. No living tissue is invaded and there is no cellular response from the host. These infections are often so innocuous that patients are often unaware of their condition.

Cutaneous Mycoses

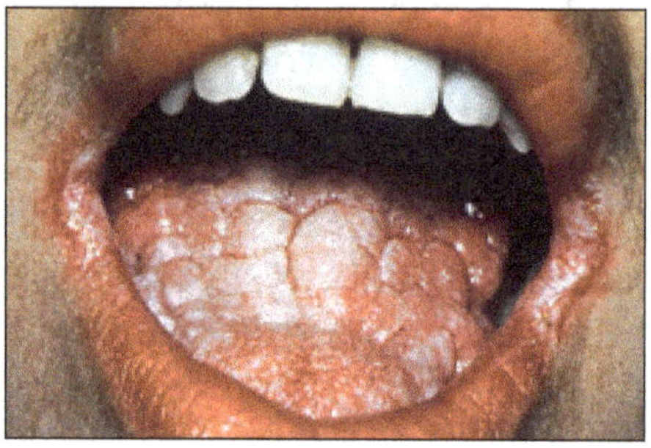

These are superficial fungal infections of the skin, hair or nails. No living tissue is invaded, however a variety of pathological changes occur in the host because of the presence of the infectious agent and its metabolic products.

Subcutaneous Mycoses

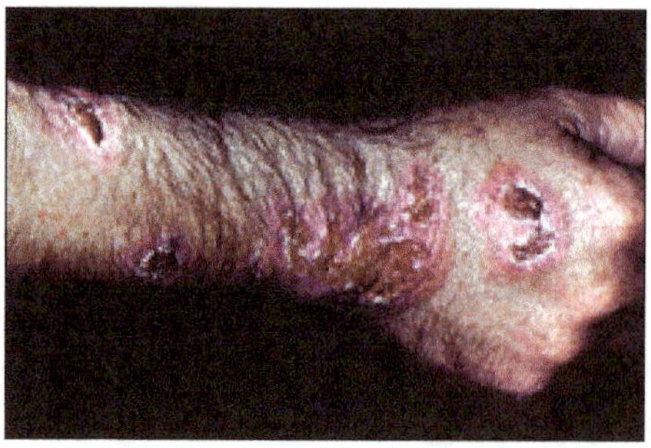

These are chronic, localized infections of the skin and subcutaneous tissue following the traumatic implantation of the aetiologic agent. The causative fungi are all soil saprophytes whose ability to adapt to the tissue environment and elicit disease is extremely variable.

Dimorphic Systemic Mycoses

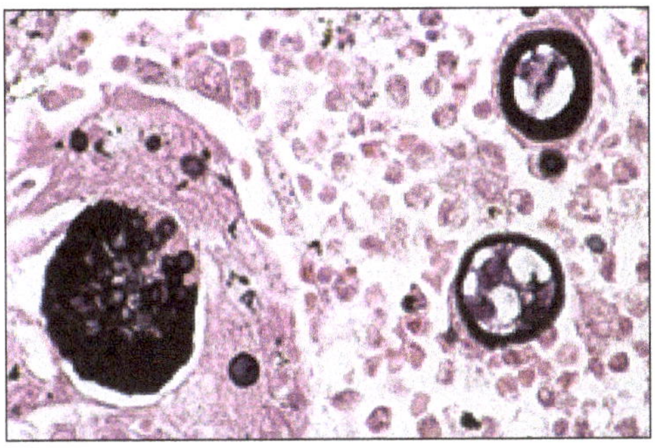

These are fungal infections caused by fungal pathogens which can overcome the defences of the normal human host by changing their morphological form. They are geographically restricted and the primary site of infection is usually pulmonary, following the inhalation of conidia.

Opportunistic Systemic Mycoses

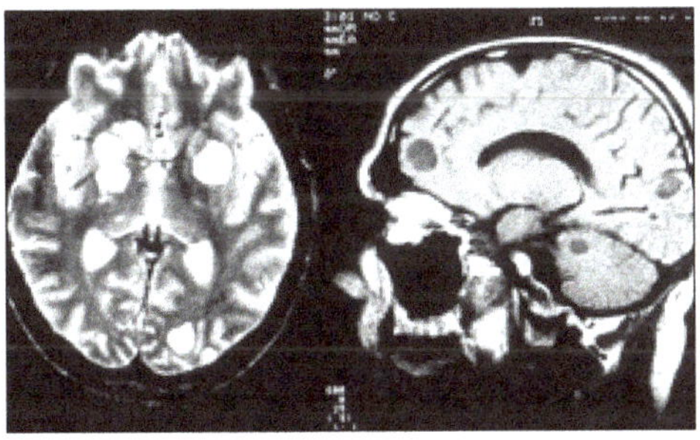

These are fungal infections which occur almost exclusively in debilitated patients whose normal defence mechanisms are impaired. The organisms involved are cosmopolitan fungi which have a very low inherent virulence.

DERMATOPHYTOSIS

Dermatophytosis, also known as ringworm, is a fungal infection of the skin. Typically it results in a red, itchy, scaly, circular rash. Hair loss may occur in the area affected. Symptoms begin four to fourteen days after exposure. Multiple areas can be affected at a given time.

About 40 types of fungi can cause ringworm. They are typically of the *Trichophyton*, *Microsporum*, or *Epidermophyton* type. Risk factors include using public showers, contact sports such as wrestling, excessive sweating, contact with animals, obesity, and poor immune function. Ringworm can spread from other animals or between people. Diagnosis is often based on the appearance and symptoms. It may be confirmed by either culturing or looking at a skin scraping under a microscope.

Prevention is by keeping the skin dry, not walking barefoot in public, and not sharing personal items. Treatment is typically with antifungal creams such as clotrimazole or miconazole. If the scalp is involved, antifungals by mouth such as fluconazole may be needed.

Globally, up to 20% of the population may be infected by ringworm at any given time. Infections of the groin are more common in males, while infections of the scalp and body occur equally in both sexes. Infections of the scalp are most common in children while infections of the groin are most common in the elderly. Descriptions of ringworm date back to ancient history.

Signs and Symptoms

Infections on the body may give rise to typical enlarging raised red rings of ringworm. Infection on the skin of the feet may cause athlete's foot and in the groin, jock itch. Involvement of the nails is termed onychomycosis, and they may thicken, discolour, and finally crumble and fall off. They are common in most adult people, with up to 20% of the population having one of these infections at any given moment.

Animals such as dogs and cats can also be affected by ringworm, and the disease can be transmitted between animals and humans, making it a zoonotic disease.

Specific signs can be:

- Red, scaly, itchy or raised patches.
- Patches may be redder on outside edges or resemble a ring.
- Patches that begin to ooze or develop a blister.
- Bald patches may develop when the scalp is affected.
- Nails may thicken, discolour or begin to crack.

Causes

Fungi thrive in moist, warm areas, such as locker rooms, tanning beds, swimming pools, and skin folds; accordingly, those that cause dermatophytosis may be spread by using exercise machines that have not been disinfected after use, or by sharing towels, clothing, or footwear, such as rental bowling shoes, ski boots, ice skates or roller skates.

Diagnosis

Classification

A number of different species of fungi are involved in dermatophytosis. Dermatophytes of the genera *Trichophyton* and *Microsporum* are the most common causative agents. These fungi attack various parts of the body and lead to the conditions listed below. The Latin names are for the conditions (disease patterns), not the agents that cause them. The disease patterns below identify the type of fungus that causes them only in the cases listed:

- Dermatophytosis:
  - Tinea pedis (athlete's foot): Fungal infection of the feet.
  - Tinea unguium: Fungal infection of the fingernails and toenails, and the nail bed.
  - Tinea corporis: Fungal infection of the arms, legs, and trunk.
  - Tinea cruris (jock itch): Fungal infection of the groin area.
  - Tinea manuum: Fungal infection of the hands and palm area.
  - Tinea capitis: Fungal infection of the scalp and hair.
  - Tinea faciei (face fungus): Fungal infection of the face.
  - Tinea barbae: Fungal infestation of facial hair.
- Other superficial mycoses (not classic ringworm, since not caused by dermatophytes):
  - Tinea versicolor: Caused by *Malassezia furfur*.
  - Tinea nigra: Caused by *Hortaea werneckii*.

Prevention

Advice often given includes:

- Avoid sharing clothing, sports equipment, towels, or sheets.
- Wash clothes in hot water with fungicidal soap after suspected exposure to ringworm.
- Avoid walking barefoot; instead wear appropriate protective shoes in locker rooms and sandals at the beach.
- Avoid touching pets with bald spots, as they are often carriers of the fungus.

Vaccination

As of 2016, no approved human vaccine exist against *Dermatophytosis*. For horses, dogs and cats there is available an approved inactivated vaccine called *Insol Dermatophyton* (Boehringer Ingelheim) which provides time-limited protection against several trichophyton and microsporum fungal strains. With cattle, systemic vaccination has achieved effective control of ringworm. Since 1979 a Russian live vaccine (LFT 130) and later on a Czechoslovakian live vaccine against bovine ringworm has been used. In Scandinavian countries vaccination programmes against ringworm are used as a preventive measure to improve the hide quality. In Russia, fur-bearing animals (silver fox, foxes, polar foxes) and rabbits have also been treated with vaccines.

Treatment

Antifungal treatments include topical agents such as miconazole, terbinafine, clotrimazole, ketoconazole, or tolnaftate applied twice daily until symptoms resolve — usually within one or two weeks. Topical treatments should then be continued for a further 7 days after resolution of visible symptoms to prevent recurrence. The total duration of treatment is therefore generally two weeks, but may be as long as three.

In more severe cases or scalp ringworm, systemic treatment with oral medications may be given.

To prevent spreading the infection, lesions should not be touched, and good hygiene maintained with washing of hands and the body.

Misdiagnosis and treatment of ringworm with a topical steroid, a standard treatment of the superficially similar pityriasis rosea, can result in tinea incognito, a condition where ringworm fungus grows without typical features, such as a distinctive raised border.

Other Animals

Ringworm caused by *Trichophyton verrucosum* is a frequent clinical condition in cattle. Young animals are more frequently affected. The lesions are located on the head, neck, tail, and perineum. The typical lesion is a round, whitish crust. Multiple lesions may coalesce in "map-like" appearance.

Multiple lesions, head.

On neck and withers..

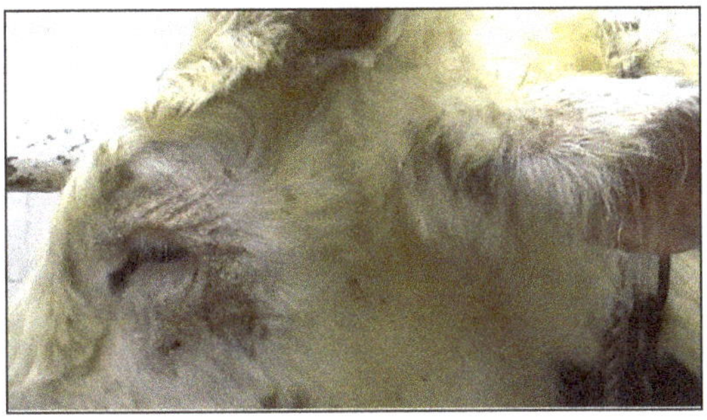

Around the eyes and on ears.

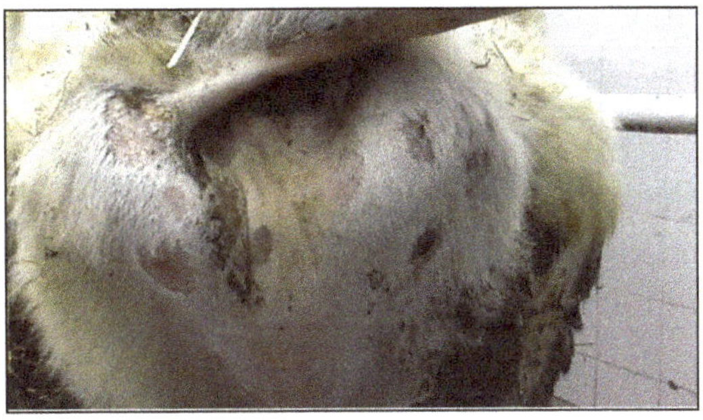

On perineum.

Clinical dermatophytosis is also diagnosed in sheep, dogs, cats, and horses. Causative agents, besides *Trichophyton verrucosum, are T. mentagrophytes, T. equinum, Microsporum gypseum, M. canis*, and *M. nanum*.

Dermatophytosis may also be present in the holotype of the Cretaceous eutriconodont mammal *Spinolestes*, suggesting a Mesozoic origin for this disease.

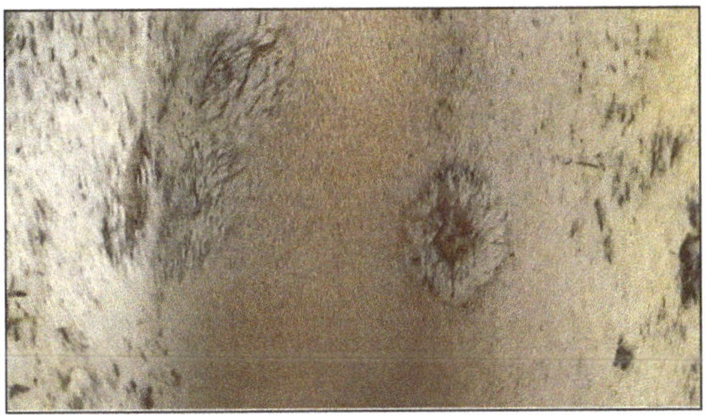

Old lesions, with regrowing hair.

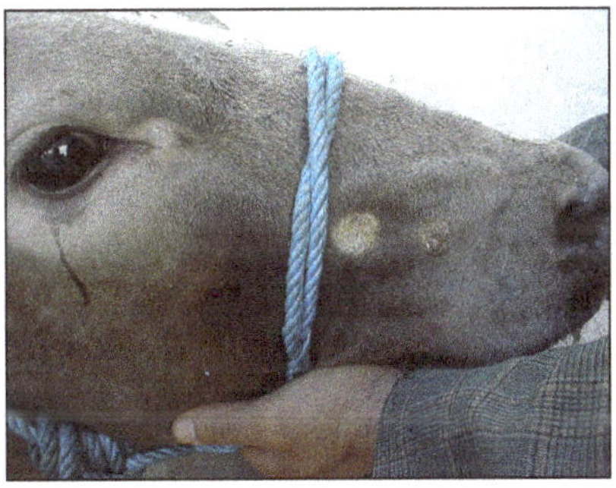

On cheeks: crusted lesion (right)

Diagnosis

Ringworm in pets may often be asymptomatic, resulting in a carrier condition which infects other pets. In some cases, the disease only appears when the animal develops an immunodeficiency condition. Circular bare patches on the skin suggest the diagnosis, but no lesion is truly specific to the fungus. Similar patches may result from allergies, sarcoptic mange, and other conditions. Three species of fungi cause 95% of dermatophytosis in pets: These are *Microsporum canis, Microsporum gypseum*, and *Trichophyton mentagrophytes*.

Veterinarians have several tests to identify ringworm infection and identify the fungal species that cause it:

Woods test: This is an ultraviolet light with a magnifying lens. Only 50% of *M. canis* will show up as an apple-green fluorescence on hair shafts, under the UV light. The other fungi do not show. The fluorescent material is not the fungus itself (which does not fluoresce), but rather an excretory product of the fungus which sticks to hairs. Infected skin does not fluoresce.

Microscopic test: The veterinarian takes hairs from around the infected area and places them in a staining solution to view under the microscope. Fungal spores may be viewed directly on hair shafts. This technique identifies a fungal infection in about 40%–70% of the infections, but cannot identify the species of dermatophyte.

Culture test: This is the most effective, but also the most time-consuming, way to determine if ringworm is on a pet. In this test, the veterinarian collects hairs from the pet, or else collects fungal spores from the pet's hair with a toothbrush, or other instrument, and inoculates fungal media for culture. These cultures can be brushed with transparent tape and then read by the veterinarian using a microscope, or can be sent to a pathological lab. The three common types of fungi which commonly cause pet ringworm can be identified by their characteristic spores. These are different-appearing macroconidia in the two common species of *Microspora*, and typical microconidia in *Trichophyton* infections.

Identifying the species of fungi involved in pet infections can be helpful in controlling the source of infection. *M. canis*, despite its name, occurs more commonly in domestic cats, and 98% of cat infections are with this organism. It can also infect dogs and humans, however. *T. mentagrophytes* has a major reservoir in rodents, but can also infect pet rabbits, dogs, and horses. *M. gypseum* is a soil organism and is often contracted from gardens and other such places. Besides humans, it may infect rodents, dogs, cats, horses, cattle, and swine.

Treatment

Pet Animals

Treatment requires both systemic oral treatment with most of the same drugs used in humans—terbinafine, fluconazole, or itraconazole—as well as a topical "dip" therapy.

Because of the usually longer hair shafts in pets compared to those of humans, the area of infection and possibly all of the longer hair of the pet must be clipped to decrease the load of fungal spores clinging to the pet's hair shafts. However, close shaving is usually not done because nicking the skin facilitates further skin infection.

Twice-weekly bathing of the pet with diluted lime sulfur dip solution is effective in eradicating fungal spores. This must continue for 3 to 8 weeks.

Washing of household hard surfaces with 1:10 household sodium hypochlorite bleach solution is effective in killing spores, but it is too irritating to be used directly on hair and skin.

Pet hair must be rigorously removed from all household surfaces, and then the vacuum cleaner bag, and perhaps even the vacuum cleaner itself, discarded when this has been done repeatedly. Removal of all hair is important, since spores may survive 12 months or even as long as two years on hair clinging to surfaces.

Cattle

In bovines, an infestation is difficult to cure, as systemic treatment is uneconomical. Local treatment with iodine compounds is time-consuming, as it needs scraping of crusty lesions. Moreover, it must be carefully conducted using gloves, lest the worker become infested.

FUNGAL KERATITIS

A fungal keratitis is an 'inflammation of the eye's cornea' (called *keratitis*) that results from infection by a fungal organism. Keratomycosis is the Greek terminology equivalent of fungal keratitis - it is the fungal infection of the cornea, the anterior part of the eye which covers the pupil. Those experiencing these symptoms are typically advised to immediately visit the appropriate eyecare professional.

Presentation

The symptoms of fungal keratitis are blurred vision, a red and painful eye that does not improve when contact lenses are removed, or on antibiotic treatment, increased sensitivity to light (photophobia), and excessive tearing or discharge. The symptoms are markedly less as compared to a similar bacterial ulcer.

Signs: The eyelids and adnexa involved shows edema and redness, conjuctiva is chemosed. Ulcer may be present. It is a dry looking corneal ulcer with satellite lesions in the surrounding cornea. Usually associated with fungal ulcer is hypopyon, which is mostly white fluffy in appearance. Rarely, it may extend to the posterior segment to cause endophthalmitis in later stages, leading to the destruction of the eye.

Causes

Filamentous fungi:

- Aspergillus flavus.
- Aspergillus fumigatus.
- Fusarium spp.
- Alternaria spp.

- Curvularia.
- Acremonium.

Yeasts:

- Candida.

Pathophysiology

The precipitating event for fungal keratitis is trauma with a vegetable / organic matter. A thorn injury, or in agriculture workers, trauma with a wheat plant while cutting the harvest is typical. This implants the fungus directly in the cornea. The fungus grows slowly in the cornea and proliferates to involve the anterior and posterior stromal layers. The fungus can break through the descemet's membrane and pass into the anterior chamber. The patient presents a few days or weeks later with fungal keratitis.

Diagnosis

The diagnosis is made by an ophthalmologist/optometrist correlating typical history, symptoms and signs. Many times it may be missed and misdiagnosed as bacterial ulcer. A definitive diagnosis is established only after a positive culture report (lactophenol cotton blue, calcoflour medium), typically taking a week, from the corneal scraping. Recent advances have been made in PCR ref 3./immunologic tests which can give a much quicker result.

Classification

Infectious keratitis can be bacterial, fungal, viral, or protozoal. Remarkable differences in presentation of the patient allows presumptive diagnosis by the eye care professional, helping in institution of appropriate anti-infective therapy.

Prevention

Prevention of trauma with vegetable/organic matter, particularly in agricultural workers while harvesting can reduce the incidence of fungal keratitis. Wearing of broad protective glasses with side shields is recommended for people at risk for such injuries.

Treatment

A presumptive diagnosis of fungal keratitis requires immediate empirical therapy. Natamycin ophthalmic suspension is the drug of choice for filamentous fungal infection. Fluconazole ophthalmic solution is recommended for Candida infection of the cornea. Amphotericin B eye drops may be required for non-responding cases, but can be quite toxic and requires expert pharmacist for preparation. Other medications have also been tried with moderate success. Consult your eye care professional in any case as they will have the best treatment.

Prognosis

The infection typically takes a long time to heal, since the fungus itself is slow growing. Corneal perforation can occur in patients with untreated or partially treated infectious keratitis and requires surgical intervention in the form of corneal transplantation.

SPOROTRICHOSIS

Sporotrichosis is a disease caused by the infection of the fungus *Sporothrix schenckii*. This fungal disease usually affects the skin, although other rare forms can affect the lungs, joints, bones, and even the brain. Because roses can spread the disease, it is one of a few diseases referred to as *rose-thorn* or *rose-gardeners' disease*. The species was named for Benjamin Schenck, a medical student who in 1896 was the first to isolate it from a human specimen.

Because *S. schenckii* is naturally found in soil, hay, sphagnum moss, and plants, it usually affects farmers, gardeners, and agricultural workers. It enters through small cuts and abrasions in the skin to cause the infection. In case of sporotrichosis affecting the lungs, the fungal spores enter through the respiratory pathways. Sporotrichosis can also be acquired from handling cats with the disease; it is an occupational hazard for veterinarians.

Sporotrichosis progresses slowly – the first symptoms may appear 1 to 12 weeks (average 3 weeks) after the initial exposure to the fungus. Serious complications can also develop in patients who have a compromised immune system.

Signs and Symptoms

- Cutaneous or skin sporotrichosis:

This is the most common form of this disease. Symptoms of this form include nodular lesions or bumps in the skin, at the point of entry and also along lymph nodes and vessels. The lesion starts off small and painless, and ranges in color from pink to purple. Left untreated, the lesion becomes larger and look similar to a boil and more lesions will appear, until a chronic ulcer develops.

Usually, cutaneous sporotrichosis lesions occur in the finger, hand, and arm.

- Pulmonary sporotrichosis:

This rare form of the disease occur when *S. schenckii* spores are inhaled. Symptoms of pulmonary sporotrichosis include productive coughing, nodules and cavitations of the lungs, fibrosis, and swollen hilar lymph nodes. Patients with this form of sporotrichosis are susceptible to developing tuberculosis and pneumonia.

- Disseminated sporotrichosis:

When the infection spreads from the primary site to secondary sites in the body, the disease develops into a rare and critical form called disseminated sporotrichosis. The infection can spread to joints and bones (called *osteoarticular sporotrichosis*) as well as the central nervous system and the brain (called *sporotrichosis meningitis*).

The symptoms of disseminated sporotrichosis include weight loss, anorexia, and appearance of bony lesions.

Complications

Cutaneous lesions can become superinfected with bacteria, resulting in cellulitis.

Diagnosis

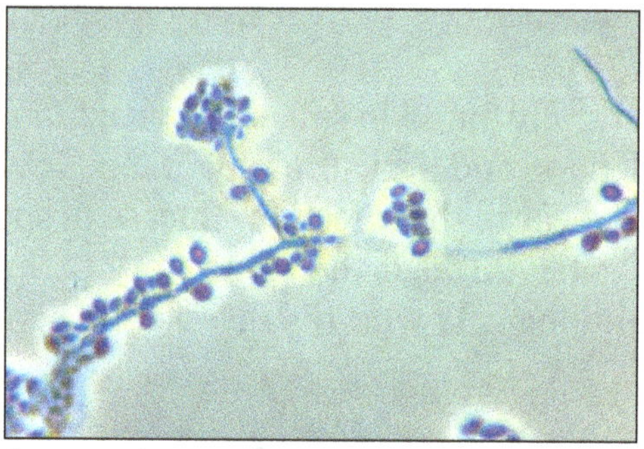

Conidiophores and conidia of the fungus *Sporothrix schenckii*.

Sporotrichosis is a chronic disease with slow progression and often subtle symptoms. It is difficult to diagnose, as many other diseases share similar symptoms and therefore must be ruled out.

Patients with sporotrichosis will have antibody against the fungus *S. schenckii*, however, due to variability in sensitivity and specificity, it may not be a reliable diagnosis for this disease. The confirming diagnosis remains culturing the fungus from the skin, sputum, synovial fluid, and cerebrospinal fluid. Smears should be taken from the draining tracts and ulcers.

Cats with sporotrichosis are unique in that the exudate from their lesions may contain numerous organisms. This makes cytological evaluation of exudate a valuable diagnostic tool in this species. Exudate is pyogranulomatous and phagocytic cells may be packed with yeast forms. These are variable in size, but many are cigar-shaped.

Prevention

The majority of sporotrichosis cases occur when the fungus is introduced through a cut or puncture in the skin while handling vegetation containing the fungal spores.

Prevention of this disease includes wearing long sleeves and gloves while working with soil, hay bales, rose bushes, pine seedlings, and sphagnum moss. Also, keeping cats indoors is a preventative measure.

Treatment

Treatment of sporotrichosis depends on the severity and location of the disease. The following are treatment options for this condition:

- Oral potassium iodide:

  Potassium iodide is an anti-fungal drug that is widely used as a treatment for cutaneous sporotrichosis. Despite its wide use, there is no high-quality evidence for or against this practice. Further studies are needed to assess the efficacy and safety of oral potassium iodide in the treatment of sporotrichosis.

- Itraconazole (Sporanox) and fluconazole:

  These are antifungal drugs. Itraconazole is currently the drug of choice and is significantly more effective than fluconazole. Fluconazole should be reserved for patients who cannot tolerate itraconazole.

- Amphotericin B:

  This antifungal medication is delivered intravenously. Many patients, however, cannot tolerate Amphotericin B due to its potential side effects of fever, nausea, and vomiting.

  Lipid formulations of amphotericin B are usually recommended instead of amphotericin B deoxycholate because of a better adverse-effect profile. Amphotericin B can be used for severe infection during pregnancy. For children with disseminated or severe disease, amphotericin B deoxycholate can be used initially, followed by itraconazole.

  In case of sporotrichosis meningitis, the patient may be given a combination of Amphotericin B and 5-fluorocytosine/Flucytosine.

- Terbinafine:

  500mg and 1000mg daily dosages of terbinafine for twelve to 24 weeks has been used to treat cutaneous sporotrichosis.

- Newer triazoles:

  Several studies have shown that posaconazole has in vitro activity similar to that of amphotericin B and itraconazole; therefore, it shows promise as an alternative therapy. However, voriconazole susceptibility varies. Because the correlation between in vitro data and clinical response has not been demonstrated, there is insufficient evidence to recommend either posaconazole or voriconazole for treatment of sporotrichosis at this time.

- Surgery:

  In cases of bone infection and cavitary nodules in the lungs, surgery may be necessary.

- Heat therapy:

  Heat creates higher tissue temperatures, which may inhibit fungus growth while the immune system counteracts the infection. The "pocket warmer" used for this purpose has the advantage of being able to maintain a constant temperature of 44 degrees-45 degrees C on the skin surface for several hours, while permitting unrestricted freedom of movement. The duration of treatment depends on the type of lesion, location, depth, and size. Generally, local application for 1-2 h per day, or in sleep time, for 5-6 weeks seems to be sufficient.

BLASTOMYCOSIS

Blastomycosis is a fungal infection caused by inhaling *Blastomyces dermatitidis* spores. If it involves only the lungs, it is called pulmonary blastomycosis. Only about half of people with the disease have symptoms, which can include fever, cough, night sweats, muscle pains, weight loss, chest pain, and feeling tired. These symptoms usually develop between three weeks and three months after breathing in the spores. In those with weak immune systems, the disease can spread to other areas of the body, including the skin and bones.

Blastomyces dermatitis is found in the soil and decaying organic matter like wood or leaves. Participating in outdoor activities like hunting or camping in wooded areas increases the risk of developing blastomycosis. There is no vaccine, but the disease can be prevented by not disturbing the soil. Treatment is with itraconazole for mild or moderate disease and amphotericin B for severe disease. With both, the duration of treatment is 6–12 months. Overall, 4-6% of people who develop blastomycosis die; however, if the central nervous system is involved, this rises to 18%. People with AIDS or on medications that suppress the immune system have the highest risk of death at 25-40%.

Blastomycosis is endemic to the eastern United States, especially the Ohio and Mississippi River valleys, the Great Lakes, and the St. Lawrence River. It is also endemic to some parts of Canada, including Quebec, Ontario, and Manitoba. In these areas, there are about 1 to 2 cases per 100,000 per year. Blastomycosis was first described by Thomas Casper Gilchrist in 1894; because of this, it is sometimes called "Gilchrist's disease".

Signs and Symptoms

The severity of symptoms depends on a person's immune status; less than 50% of healthy people with blastomycosis have symptoms, while immunocompromised patients can have the disease spread beyond the lungs to other organs like the skin and bones.

Blastomycosis can present in one of the following ways:

- A flu-like illness with fever, chills, arthralgia (joint pain), myalgia (muscle pain), headache, and a nonproductive cough which resolves within days.
- An acute illness resembling bacterial pneumonia, with symptoms of high fever, chills, a productive cough, and pleuritic chest pain.
- A chronic illness that mimics tuberculosis or lung cancer, with symptoms of low-grade fever, a productive cough, night sweats, and weight loss.
- A fast, progressive, and severe disease that manifests as ARDS, with fever, shortness of breath, tachypnea, hypoxemia, and diffuse pulmonary infiltrates.
- Skin lesions, usually asymptomatic, can be verrucous (wart-like) or ulcerated with small pustules at the margins.
- Bone lytic lesions can cause bone or joint pain.
- prostatitis may be asymptomatic or may cause pain on urinating.
- Laryngeal involvement causes hoarseness.
- 40% immunocompromised individuals have CNS involvement and present as brain abscess, epidural abscess or meningitis.

Cause

Blastomycosis is caused by the dimorphic microfungus *Blastomyces dermatitidis*, a member of the phylum Ascomycota in the family Ajellomycetaceae. It has been recognised as the asexual state of *Ajellomyces dermatitidis*. In endemic areas, the fungus lives in soil and rotten wood near lakes and rivers. Although it has never been directly observed growing in nature, it is thought to grow there as a cottony white mold, similar to the growth seen in artificial culture at 25 °C. The moist, acidic soil in the surrounding woodland harbors the fungus.

Spectrum of Disease

Blastomycosis manifests as a primary lung infection in about 70% of cases. The onset is relatively slow and symptoms are suggestive of pneumonia, often leading to initial treatment with antibacterials. Occasionally, if a lesion is seen on X-ray in a cigarette smoker, the disease may be misdiagnosed as carcinoma, leading to swift excision of the pulmonary lobe involved. Upper lung lobes are involved somewhat more frequently than lower lobes. If untreated, many cases progress over a period of months to years to become disseminated blastomycosis. In these cases, the large Blastomyces yeast cells translocate from the lungs and are trapped in capillary beds elsewhere in the body, where they cause lesions. The skin is the most common organ affected, being the site of lesions in approximately 60% of cases. The signature image of blastomycosis in textbooks is the indolent, verrucous or ulcerated dermal lesion seen in disseminated disease. Osteomyelitis is also common (12–60% of cases). Other recurring sites of dissemination are the genitourinary tract (kidney, prostate, epididymis; collectively ca. 25% of cases) and the brain (3–10% of cases).

An uncommon but very dangerous type of primary blastomycosis manifests as acute respiratory distress syndrome (ARDS); for example, this was seen in 9 of 72 blastomycosis cases studied in northeast Tennessee. Such cases may follow massive exposure, e.g., during brush clearing operations. The fatality rate in the ARDS cases in the Tennessee study was 89%, while in non-ARDS cases of pulmonary blastomycosis, the fatality rate was 10%.

Pathogenesis

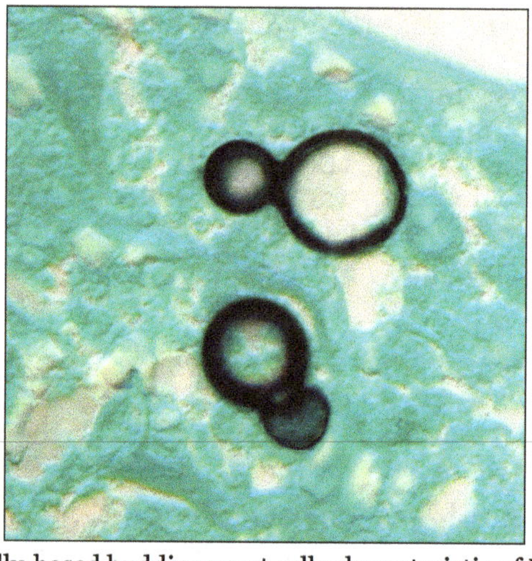

Large, broadly-based budding yeast cells characteristic of Blastomyces dermatitidis in a GMS-stained biopsy section from a human leg.

Inhaled conidia of *B. dermatitidis* are phagocytosed by neutrophils and macrophages in alveoli. Some of these escape phagocytosis and transform into yeast phase rapidly. Having thick walls, these are resistant to phagocytosis and express glycoprotein, *BAD-1*, which is a virulence factor as well as an epitope. In lung tissue, they multiply and may disseminate through blood and lymphatics to other organs, including the skin, bone, genitourinary tract, and brain. The incubation period is 30 to 100 days, although infection can be asymptomatic.

Diagnosis

Once suspected, the diagnosis of blastomycosis can usually be confirmed by demonstration of the characteristic broad based budding organisms in sputum or tissues by KOH prep, cytology, or histology. Tissue biopsy of skin or other organs may be required in order to diagnose extra-pulmonary disease. Blastomycosis is histologically associated with granulomatous nodules. Commercially available urine antigen testing appears to be quite sensitive in suggesting the diagnosis in cases where the organism is not readily detected. While culture of the organism remains the definitive diagnostic standard, its slow growing nature can lead to delays in treatment of up to several weeks. However, sometimes blood and sputum cultures may not detect blastomycosis.

Treatment

Itraconazole given orally is the treatment of choice for most forms of the disease. Ketoconazole may also be used. Cure rates are high, and the treatment over a period of months is usually well tolerated. Amphotericin B is considerably more toxic, and is usually reserved for immunocompromised patients who are critically ill and those with central nervous system disease. Patients who cannot tolerate deoxycholate formulation of Amphotericin B can be given lipid formulations. Fluconazole has excellent CNS penetration and is useful where there is CNS involvement after initial treatment with Amphotericin B.

Prognosis

Mortality rate in treated cases:

- 0-2% in treated cases among immunocompetent patients.
- 29% in immunocompromised patients.
- 40% in the subgroup of patients with AIDS.
- 68% in patients presenting as acute respiratory distress syndrome (ARDS).

HISTOPLASMOSIS

Histoplasmosis is a disease caused by the fungus *Histoplasma capsulatum*. Symptoms of this infection vary greatly, but the disease affects primarily the lungs. Occasionally, other organs are affected; this is called disseminated histoplasmosis, and it can be fatal if left untreated.

Histoplasmosis is common among AIDS patients because of their suppressed immunity. In immunocompetent individuals, past infection results in partial protection against ill effects if reinfected.

Histoplasma capsulatum is found in soil, often associated with decaying bat guano or bird droppings. Disruption of soil from excavation or construction can release infectious elements that are inhaled and settle into the lung.

Signs and Symptoms

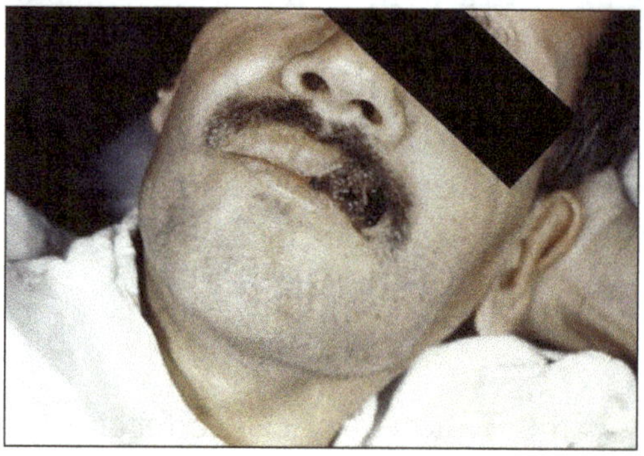

Skin lesion on the upper lip due to Histoplasma capsulatum infection.

If symptoms of histoplasmosis infection occur, they will start within 3 to 17 days after exposure; the average is 12–14 days. Most affected individuals have clinically silent manifestations and show no apparent ill effects. The acute phase of histoplasmosis is characterized by non-specific respiratory symptoms, often cough or flu-like. Chest X-ray findings are normal in 40–70% of cases. Chronic histoplasmosis cases can resemble tuberculosis; disseminated histoplasmosis affects multiple organ systems and is fatal unless treated.

While histoplasmosis is the most common cause of mediastinitis, this remains a relatively rare disease. Severe infections can cause hepatosplenomegaly, lymphadenopathy, and adrenal enlargement. Lesions have a tendency to calcify as they heal.

Presumed ocular histoplasmosis syndrome (POHS) causes chorioretinitis, where the choroid and retina of the eyes are scarred, resulting in a loss of vision not unlike macular degeneration. Despite its name, the relationship to *Histoplasma* is controversial. Distinct from POHS, acute ocular histoplasmosis may rarely occur in immunodeficiency.

Complications

In absence of proper treatment and especially in immunocompromised individuals, complications can arise. These include recurrent pneumonia, respiratory failure, fibrosing mediastinitis, superior vena cava syndrome, pulmonary vessel obstruction, progressive fibrosis of lymph nodes. Fibrosing mediastinitis is a serious complication and can be fatal. Smokers with structural lung disease have higher probability of developing chronic cavitary histoplasmosis.

After healing of lesions, hard calcified lymph nodes can erode the walls of airway causing hemoptysis.

Mechanisms

H. capsulatum grows in soil and material contaminated with bird or bat droppings (guano). The fungus has been found in poultry house litter, caves, areas harboring bats, and in bird roosts (particularly those of starlings). The fungus is thermally dimorphic: in the environment it grows as a brownish mycelium, and at body temperature (37 °C in humans) it morphs into a yeast. Histoplasmosis is not contagious, but is contracted by inhalation of the spores from disturbed soil or guano. The inoculum is represented principally by microconidia. These are inhaled and reach the alveoli.

In the alveoli, macrophages ingest these microconidia. They survive inside the phagosome. As the fungus is thermally dimorphic, these microconidia are transformed into yeast. They grow and multiply inside the phagosome. The macrophages travel in lymphatic circulation and spread the disease to different organs.

Within the phagosome the fungus has an absolute requirement for thiamine. Cell-mediated immunity for histoplasmosis develops within 2 weeks. If the patient has strong cellular immunity, macrophages, epithelial cells and lymphocytes surround the organisms and contain them, and eventually calcify. In immunocompromised individuals, the organisms disseminate to different organs such as bone, spleen, liver, adrenal glands and mucocutaneous membranes, resulting in progressive disseminated histoplasmosis. Chronic lung disease can manifest.

Diagnosis

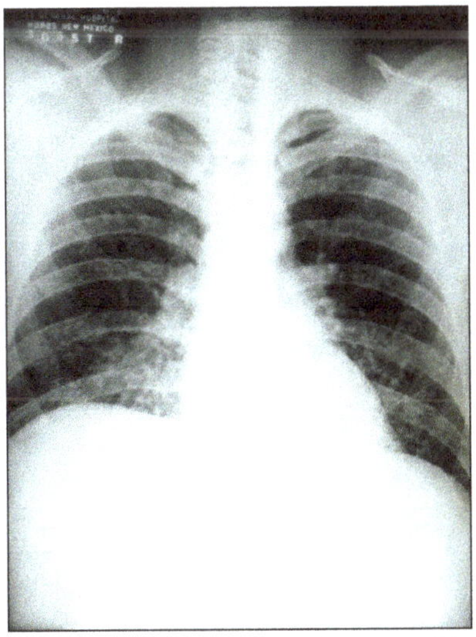

Chest X-ray of a patient with acute pulmonary histoplasmosis.

Clinically, there is a wide spectrum of disease manifestation, making diagnosis somewhat difficult. More severe forms include: (1) The chronic pulmonary form, often occurring in the presence of underlying pulmonary disease; and (2) a disseminated form, which is characterized by the progressive spread of infection to extra-pulmonary sites. Oral manifestations have been reported as the main complaint of the disseminated forms, leading the patient to seek treatment, whereas pulmonary symptoms in disseminated disease may be mild or even misinterpreted as flu. Histoplasmosis can be diagnosed by samples containing the fungus taken from sputum (via bronchoalveolar lavage), blood, or infected organs. It can also be diagnosed by detection of antigens in blood or urine samples by ELISA or PCR. Antigens can cross-react with antigens of African histoplasmosis (caused by Histoplasma duboisii), blastomycosis, coccidioidomycosis, paracoccidioidomycosis, and talaromycosis infection. Histoplasmosis can also be diagnosed by a test for antibodies against *Histoplasma* in the blood. *Histoplasma* skin tests indicate whether a person has been exposed, but do not indicate whether they have the disease. Formal histoplasmosis diagnoses are often confirmed only by culturing the fungus directly. Sabouraud agar is one type of agar growth media on which the fungus can be cultured. Cutaneous manifestations of disseminated disease are diverse and often present as a nondescript rash with systemic complaints. Diagnosis is best established by urine antigen testing, as blood cultures may take up to 6 weeks for diagnostic growth to occur and serum antigen testing often comes back with a false negative before 4 weeks of disseminated infection.

Types

Histoplasmosis may be divided into the following types:

- Primary pulmonary histoplasmosis.
- Progressive disseminated histoplasmosis.
- Primary cutaneous histoplasmosis.
- African histoplasmosis.

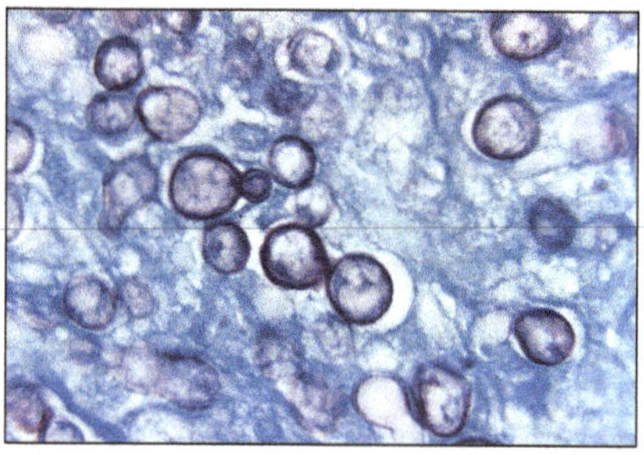

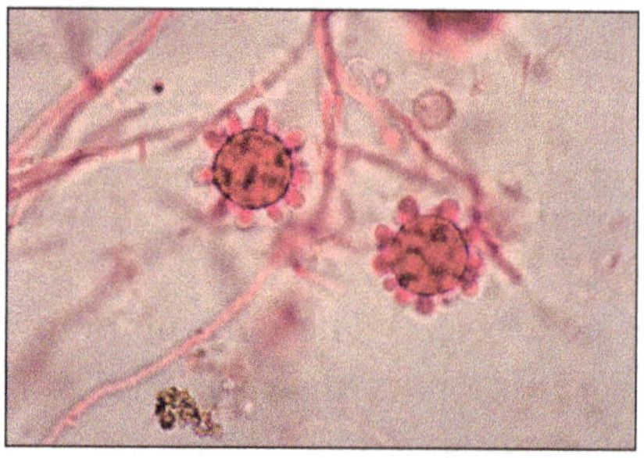

Prevention

It is not practical to test or decontaminate most sites that may be contaminated with *H. capsulatum*, but the following sources list environments where histoplasmosis is common, and precautions to reduce a person's risk of exposure, in the three parts of the world where the disease is prevalent. Precautions common to all geographical locations would be to avoid accumulations of bird or bat droppings.

The US National Institute for Occupational Safety and Health (NIOSH) provides information on work practices and personal protective equipment that may reduce the risk of infection. This document is available in English and Spanish.

Treatment

In the majority of immunocompetent individuals, histoplasmosis resolves without any treatment. Antifungal medications are used to treat severe cases of acute histoplasmosis and all cases of chronic and disseminated disease. Typical treatment of severe disease first involves treatment with amphotericin B, followed by oral itraconazole.

Liposomal preparations of amphotericin B are more effective than deoxycholate preparations. The liposomal preparation is preferred in patients that might be at risk of nephrotoxicity, although all preparations of amphotericin B have risk of nephrotoxicity. Individuals taking amphotericin B are monitored for renal function.

Treatment with itraconazole will need to continue for at least a year in severe cases, while in acute pulmonary histoplasmosis, 6 to 12 weeks treatment is sufficient. Alternatives to itraconazole are posaconazole, voriconazole, and fluconazole. Individuals taking itraconazole are monitored for hepatic function.

ASPERGILLOSIS

Aspergillosis is the name given to a wide variety of diseases caused by fungal infections from species of *Aspergillus*. Aspergillosis occurs in humans, birds and other animals.

Aspergillosis occurs in chronic or acute forms which are clinically very distinct. Most cases of acute aspergillosis occur in people with severely compromised immune systems, e.g. those undergoing bone marrow transplantation. Chronic colonization or infection can cause complications in people with underlying respiratory illnesses, such as asthma, cystic fibrosis, sarcoidosis, tuberculosis, or chronic obstructive pulmonary disease. Most commonly, aspergillosis occurs in the form of chronic pulmonary aspergillosis (CPA), aspergilloma, or allergic bronchopulmonary aspergillosis (ABPA). Some forms are intertwined; for example ABPA and simple aspergilloma can progress to CPA.

Other, noninvasive manifestations include fungal sinusitis (both allergic in nature and with established fungal balls), otomycosis (ear infection), keratitis (eye infection), and onychomycosis (nail infection). In most instances, these are less severe, and curable with effective antifungal treatment.

The most frequently identified pathogens are *Aspergillus fumigatus* and *Aspergillus flavus*, ubiquitous organisms capable of living under extensive environmental stress. Most people are thought to inhale thousands of *Aspergillus* spores daily but without effect due to an efficient immune response. Taken together, the major chronic, invasive, and allergic forms of aspergillosis account for around 600,000 deaths annually worldwide.

Symptoms

A fungus ball in the lungs may cause no symptoms and may be discovered only with a chest X-ray, or it may cause repeated coughing up of blood, chest pain, and occasionally severe, even fatal, bleeding. A rapidly invasive *Aspergillus* infection in the lungs often causes cough, fever, chest pain, and difficulty breathing.

Poorly controlled aspergillosis can disseminate through the blood stream to cause widespread organ damage. Symptoms include fever, chills, shock, delirium, seizures, and blood clots. The person may develop kidney failure, liver failure (causing jaundice), and breathing difficulties. Death can occur quickly.

Aspergillosis of the ear canal causes itching and occasionally pain. Fluid draining overnight from the ear may leave a stain on the pillow. Aspergillosis of the sinuses causes a feeling of congestion and sometimes pain or discharge. It can extend beyond the sinuses.

In addition to the symptoms, an X-ray or computerised tomography (CT) scan of the infected area provides clues for making the diagnosis. Whenever possible, a doctor sends a sample of infected material to a laboratory to confirm identification of the fungus.

Diagnosis

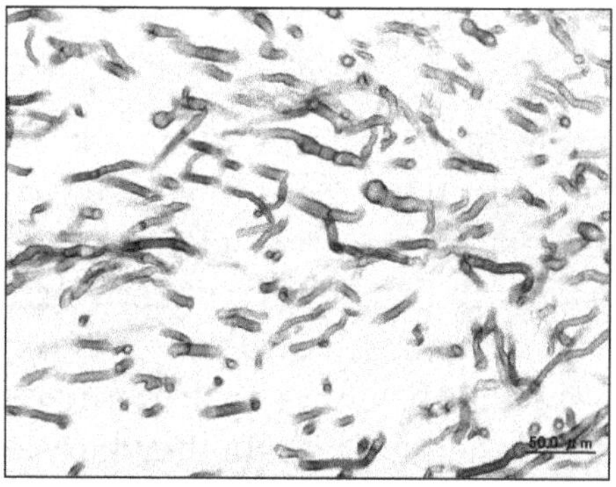

Angioinvasive pulmonary aspergillosis (closeup).

On chest X-ray and CT, pulmonary aspergillosis classically manifests as a halo sign, and later, an air crescent sign. In hematologic patients with invasive aspergillosis, the galactomannan test can make the diagnosis in a noninvasive way. False-positive *Aspergillus* galactomannan tests have been found in patients on intravenous treatment with some antibiotics or fluids containing gluconate or citric acid such as some transfusion platelets, parenteral nutrition, or PlasmaLyte.

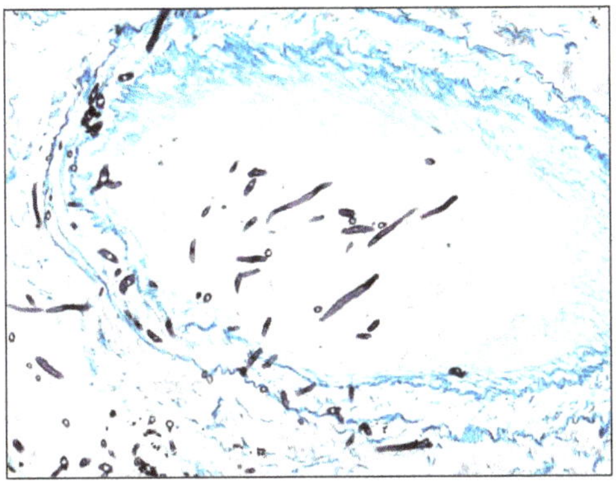

Angioinvasive pulmonary aspergillosis.

On microscopy, *Aspergillus* species are reliably demonstrated by silver stains, e.g., Gridley stain or Gomori methenamine-silver. These give the fungal walls a gray-black colour. The hyphae of *Aspergillus* species range in diameter from 2.5 to 4.5 μm. They have septate hyphae, but these are not always apparent, and in such cases they may be mistaken for Zygomycota. *Aspergillus* hyphae tend to have dichotomous branching that is progressive and primarily at acute angles of around 45°.

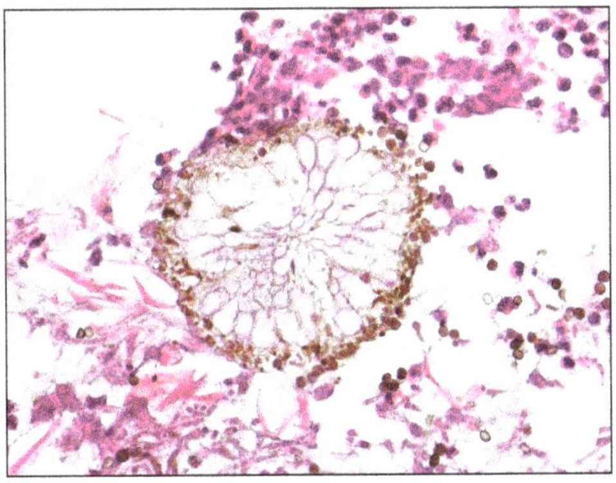

Aspergillus vesicle (HE stain).

Risk Factors

People who are immunocompromised — such as patients undergoing hematopoietic stem cell transplantation, chemotherapy for leukaemia, or AIDS — are at an increased risk for invasive aspergillosis infections. These people may have neutropenia or corticoid-induced immunosuppression as a result of medical treatments. Neutropenia is often caused by extremely cytotoxic medications such as cyclophosphamide. Cyclophosphamide interferes with cellular replication including that of white blood cells such as neutrophils. A decreased neutrophil count inhibits the ability of the body to mount immune responses against pathogens. Although tumor necrosis factor alpha (TNF-α) — a signaling molecule related to acute inflammation responses — is produced, the abnormally low number of neutrophils present in neutropenic patients leads to a depressed inflammatory response. If the underlying neutropenia is not fixed, rapid and uncontrolled hyphal growth of the invasive fungi will occur and result in negative health outcomes.

Prevention

Prevention of aspergillosis involves a reduction of mold exposure via environmental infection-control. Antifungal prophylaxis can be given to high-risk patients. Posaconazole is often given as prophylaxis in severely immunocompromised patients.

Treatment

The current medical treatments for aggressive invasive aspergillosis include voriconazole and liposomal amphotericin B in combination with surgical debridement. For the less aggressive allergic bronchopulmonary aspergillosis, findings suggest the use of oral steroids for a prolonged period of time, preferably for 6–9 months in allergic aspergillosis of the lungs. Itraconazole is given with the steroids, as it is considered to have a "steroid-sparing" effect, causing the steroids to be more effective, allowing a lower dose. Other drugs used, such as amphotericin B, caspofungin (in combination therapy only), flucytosine (in combination therapy only), or itraconazole, are used to treat this fungal infection. However, a growing proportion of infections are resistant to the triazoles. *A. fumigatus*, the most commonly infecting species, is intrinsically resistant to fluconazole.

Animals

While relatively rare in humans, aspergillosis is a common and dangerous infection in birds, particularly in pet parrots. Mallards and other ducks are particularly susceptible, as they often resort to poor food sources during bad weather. Captive raptors, such as falcons and hawks, are susceptible to this disease if they are kept in poor conditions and especially if they are fed pigeons, which are often carriers of "asper". It can be acute in chicks, but chronic in mature birds.

In the United States, aspergillosis has been the culprit in several rapid die-offs among waterfowl. From 8 December until 14 December 2006, over 2,000 mallards died near Burley, Idaho, an agricultural community about 150 miles southeast of Boise. Mouldy waste grain from the farmland and feedlots in the area is the suspected source. A similar aspergillosis outbreak caused by mouldy grain killed 500 mallards in Iowa in 2005.

While no connection has been found between aspergillosis and the H5N1 strain of avian influenza (commonly called "bird flu"), rapid die-offs caused by aspergillosis can spark fears of bird flu outbreaks. Laboratory analysis is the only way to distinguish bird flu from aspergillosis.

In dogs, aspergillosis is an uncommon disease typically affecting only the nasal passages (nasal aspergillosis). This is much more common in dolicocephalic breeds. It can also spread to the rest of the body; this is termed disseminated aspergillosis and is rare, usually affecting individuals with underlying immune disorders.

In 2019, an outbreak of aspergillosis struck the rare kakapo, a large flightless parrot endemic to New Zealand. By June the disease had killed seven of the birds, whose total population at the time was only 142 adults and 72 chicks. One fifth of the population was infected with the disease and the entire species was considered at risk of extinction.

COCCIDIOIDOMYCOSIS

Coccidioidomycosis, commonly known as cocci, Valley fever, as well as California fever, desert rheumatism, and San Joaquin Valley fever, is a mammalian fungal disease caused by *Coccidioides immitis* or *Coccidioides posadasii*. Coccidioidomycosis is endemic in certain parts of the US in Arizona, California, Nevada, New Mexico, Texas, Utah, and northern Mexico.

C. immitis is a dimorphic saprophytic fungus that grows as a mycelium in the soil and produces a spherule form in the host organism. It resides in the soil in certain parts of the southwestern United States, most notably in California and Arizona. It is also commonly found in northern Mexico, and parts of Central and South America. *C. immitis* is dormant during long dry spells, then develops as a mold with long filaments that break off into airborne spores when it rains. The spores, known as arthroconidia, are swept into the air by disruption of the soil, such as during construction, farming, or an earthquake. Windstorms may also cause epidemics far from endemic areas. In December 1977, a windstorm in an endemic area around Arvin, California led to several hundred cases, including deaths, in non-endemic areas hundreds of miles away.

Coccidioidomycosis is a common cause of community-acquired pneumonia in the endemic areas of the United States. Infections usually occur due to inhalation of the arthroconidial spores after soil disruption. The disease is not contagious. In some cases the infection may recur or become chronic.

Signs and Symptoms

An estimated 60% of people infected with the fungi responsible for coccidioidomycosis have minimal to no symptoms, while 40% will have a range of possible clinical symptoms. Of those who do develop symptoms, the primary infection is most often respiratory, with symptoms resembling bronchitis or pneumonia that resolve over a matter of a few weeks. In endemic regions, coccidioidomycosis is responsible for 20% of cases of community-acquired pneumonia. Notable coccidioidomycosis signs and symptoms include a profound feeling of tiredness, fever, cough, headaches, rash, muscle pain, and joint pain. Fatigue can persist for many months after initial infection. The classic triad of coccidioidomycosis known as "desert rheumatism" includes the combination of fever, joint pains, and erythema nodosum.

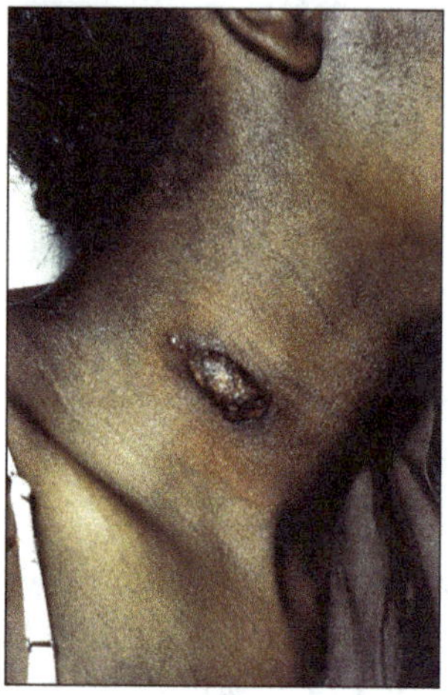

A skin lesion due to Coccidioides infection.

3% to 5% of infected individuals do not recover from the initial acute infection and develop a chronic infection. This can take the form of chronic lung infection or widespread disseminated infection (affecting the tissues lining the brain, soft tissues, joints, and bone). Chronic infection is responsible for most of the morbidity and mortality. Chronic fibrocavitary disease is manifested by cough, sputum, fevers, night sweats and weight loss. Osteomyelitis, including involvement of the spine, and meningitis may occur months to years after initial infection. Severe lung disease may develop in HIV-infected persons.

Types

After *Coccidioides* infection, coccidioidomycosis begins with Valley fever, which is its initial acute form. Valley fever may progress to the chronic form and then to disseminated coccidioidomycosis. Therefore, *Coccidioidomycosis* may be divided into the following types:

- Chronic coccidioidomycosis.
- Disseminated coccidioidomycosis, which includes primary cutaneous coccidioidomycosis.

Complications

Serious complications may occur in patients who have weakened immune systems, including severe pneumonia with respiratory failure and bronchopleural fistulas requiring resection, lung nodules, and possible disseminated form, where the infection spreads throughout the body. The disseminated form of coccidioidomycosis can devastate the body, causing skin ulcers, abscesses, bone lesions, swollen joints with severe pain, heart inflammation, urinary tract problems, and inflammation of the brain's lining, which can lead to death.

Cause

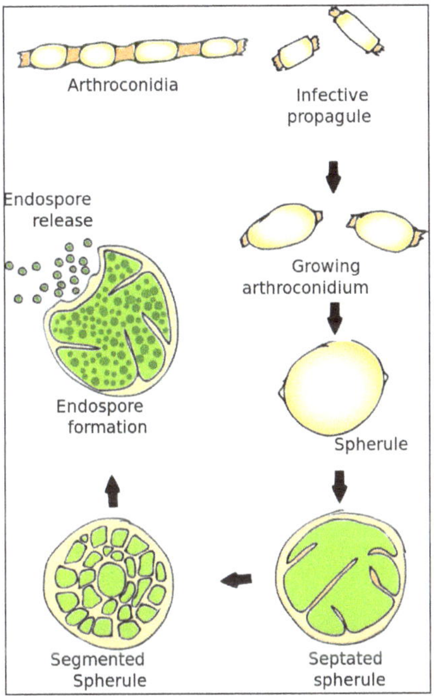

Life cycle of coccidioides.

It must rain first to start the cycle of initial growth of the fungus underneath the soil. In soil (and in agar media), *Coccidioides* exist in filament form. It forms hyphae in both horizontal and vertical directions. Over a prolonged dry period, cells within hyphae degenerate to form alternating barrel-shaped cells (arthroconidia). Arthroconidia are lightweight and carried by air currents. This happens when the soil is disturbed often by clearing trees, construction and farming. As the population grows, so have all these industries, causing a potential cascade effect. The more land that is cleared, the more arid the soil, the riper the environment for *Coccidioides*. These spores can be easily inhaled without the person knowing. On arriving in alveoli, they enlarge in size to become spherules, and internal septations develop. This division of cells is made possible by the optimal temperature inside the body. Septations develop and form endospores within the spherule. Rupture of spherules release these endospores, which in turn repeat the cycle and spread the infection to adjacent tissues within the body of the infected individual. Nodules can form in lungs surrounding these spherules. When they rupture, they release their contents into bronchi, forming thin-walled cavities. These cavities can result in symptoms like characteristic chest pain, coughing up blood, and persistent cough. In individuals with a weakened immune system, the infection can spread through the blood. On rare occasion it can enter the body through a break in the skin, causing infection.

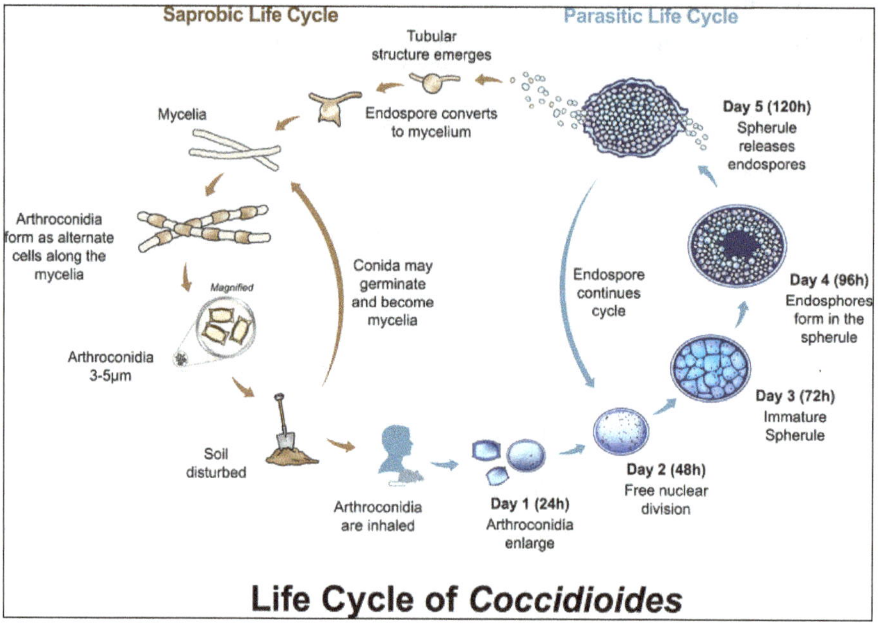

Both Coccidioides species share the same asexual life cycle, switching between saprobic and parasitic life stages.

Diagnosis

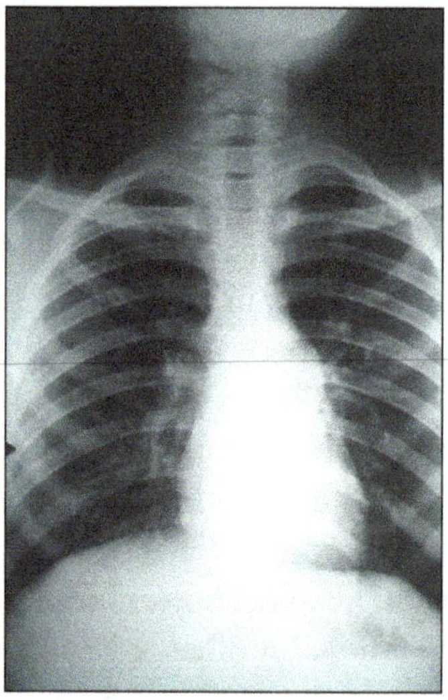
A case of pulmonary fibrosis caused by coccidioidomycosis.

Coccidioidomycosis diagnosis relies on a combination of an infected person's signs and symptoms, findings on radiographic imaging, and laboratory results. The disease is commonly misdiagnosed as bacterial community-acquired pneumonia. The fungal infection can be demonstrated by microscopic detection of diagnostic cells in body fluids, exudates, sputum and biopsy tissue by methods of Papanicolaou or Grocott's methenamine silver staining. These stains can demonstrate spherules and surrounding inflammation.

With specific nucleotide primers, *C.immitis* DNA can be amplified by polymerase chain reaction (PCR). It can also be detected in culture by morphological identification or by using molecular probes that hybridize with *C.immitis* RNA. *C. immitis* and *C. posadasii* cannot be distinguished on cytology or by symptoms, but only by DNA PCR.

An indirect demonstration of fungal infection can be achieved also by serologic analysis detecting fungal antigen or host IgM or IgG antibody produced against the fungus. The available tests include the tube-precipitin (TP) assays, complement fixation assays, and enzyme immunoassays. TP antibody is not found in cerebrospinal fluid (CSF). TP antibody is specific and is used as a confirmatory test, whereas ELISA is sensitive and thus used for initial testing.

If the meninges are affected, CSF will show abnormally low glucose levels in CSF, an increased level of protein in the CSF, and lymphocytic pleocytosis. Rarely, CSF eosinophilia is present.

Imaging

Chest x-rays rarely demonstrate nodules or cavities in the lungs, but these images commonly demonstrate lung opacification, pleural effusions, or enlargement of lymph nodes associated with the lungs. Computed tomography scans of the chest are better able to detect these changes than chest x-rays.

Prevention

Preventing Valley fever is challenging because it is difficult to avoid breathing in the fungus should it be present, however, the public health effect of the disease is essential to understand in areas where the fungus is endemic. Enhancing surveillance of Coccidiodomycosis is key to preparedness in the medical field in addition to improving diagnostics for early infections. Currently there are no completely effective preventive measures available for people who live or travel through Valley Fever -endemic areas. Recommended preventive measures include avoiding airborne dust or dirt, but this does not guarantee protection against infection. People in certain occupations may be advised to wear face masks. The use of air filtration indoors is also helpful, in addition to keeping skin injuries clean and covered to avoid skin infection.

In 1998-2011, there were 111,117 cases of coccidioidomycosis in the U.S. that were logged into the National Notifiable Diseases Surveillance System (NNDSS). Since many U.S. states do not require reporting of coccidioidomycosis, the actual numbers can be higher. The United States' Centers for Disease Control and Prevention (CDC) called the disease a "silent epidemic" and acknowledged that there is no proven anticoccidioidal vaccine available. Studies done in the past show that the cost benefit of a vaccine is most notable among infants, teens, and immigrant adults, with negative cost-benefit results among older age groups.

Raising both surveillance and awareness of the disease while medical researchers are developing a human vaccine can positively contribute towards prevention efforts. Research demonstrates that patients from endemic areas who are aware of the disease are most likely to request diagnostic testing for coccidioidomycosis. Presently, Meridian Bioscience manufactures the so-called *EIA test* to diagnose the Valley fever, which however is known for producing a fair quantity of false positives. Currently, recommended prevention measures can include type-of-exposure-based respirator protection for persons engaged in agriculture, construction and others working outdoors in endemic areas. Dust control measures such as planting grass and wetting the soil, and also limiting exposure to dust storms are advisable for residential areas in endemic regions.

Treatment

Significant disease develops in fewer than 5% of those infected and typically occurs in those with a weakened immune system. Mild asymptomatic cases often do not require any treatment, and the symptoms will go away within a few months. Those with severe symptoms may benefit from anti-fungal therapy, which usually requires 3–6 months of treatment. There is a lack of prospective studies that examine optimal anti-fungal therapy for coccidioidomycosis.

On the whole, oral fluconazole and intravenous amphotericin B are used in progressive or disseminated disease, or in immunocompromised individuals. Amphotericin B used to be the only available treatment, although now there are alternatives, including itraconazole or ketoconazole may be used for milder disease. Fluconazole is the preferred medication for coccidioidal meningitis, due to its penetration into CSF. Intrathecal or intraventricular amphotericin B therapy is used if infection persists after fluconazole treatment. Itraconazole is used for cases that involve treatment of infected person's bones and joints. The antifungal medications posaconazole and voriconazole have also been used to treat coccidioidomycosis. Because the symptoms of valley fever are similar to the common flu and other respiratory diseases, it is important for public health professionals to be aware of the rise of valley fever and the specifics of diagnosis. Greyhound dogs often get valley fever as well, and their treatment regimen involves 6–12 months of ketoconazole, to be taken with food.

Toxicity

Conventional *amphotericin B desoxycholate* (AmB: used since the 1950s as a primary agent) is known to be associated with increased drug-induced nephrotoxicity (kidney toxicity) impairing kidney function. Other formulations have been developed such as lipid soluble formulations to mitigate such side-effects as direct proximal and distal tubular cytotoxicity. These include liposomal amphotericin B, *amphotericin B lipid complex* such as Abelcet (brand) *amphotericin B phospholipid complex* also as *AmBisome Intravenous*, or *Amphotec Intravenous* (Generic; Amphotericin B Cholesteryl Sul) and, *amphotericin B colloidal dispersion*, all shown to exhibit a decrease in nephrotoxicity. The later was not as effective in one study as *amphotericin B desoxycholate* which had a 50% murine morbidity rate versus zero for the AmB colloidal dispersion.

The cost of AmB deoxycholate, in 2015, for a patient of 70 kilograms (150 lb) at 1 mg/kg/day dosage, was approximately $63.80, compared to 5 mg/kg/day of liposomal AmB at $1318.80, making the less toxic option less accessible.

PARACOCCIDIOIDOMYCOSIS

Paracoccidioidomycosis (PCM) is an acute to chronic fungal infection caused by fungi in the genus *Paracoccidioides*, including *Paracoccidioides brasiliensis* and *Paracoccidioides lutzii*. It is endemic to Central and South America, and is considered to be a neglected endemic mycosis, a type of neglected tropical disease.

Signs and Symptoms

Asymptomatic lung infection is common, with fewer than 5% of infected individuals developing clinical disease. Two presentations are known, firstly the acute or subacute form, which predominantly affects children and young adults, and the chronic form, predominantly affecting adult men. Most patients are infected before age 20, although symptoms may present many years later.

Juvenile (Acute/Subacute) Form

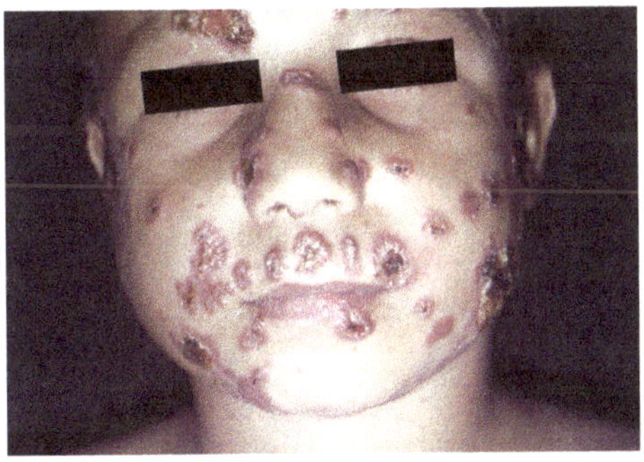

Lesions due to Paracoccidioidomycosis on the face of a Brazilian child.

The juvenile, acute form is characterised by symptoms, such as fever, weight loss and feeling unwell together with enlarged lymph nodes and enlargement of the liver and spleen. This form is most often disseminated, with symptoms manifesting depending on the organs involved. Skin and mucous membrane lesions are often present, and bone involvement may occur in severe cases. This acute, severe presentation may mimic tuberculosis, lymphoma or leukaemia.

Adult (Chronic) Form

The chronic form presents months to years after the initial infection occurs and most frequently presents with dry cough and shortness of breath. Other symptoms include excess salivation, difficulty swallowing, and difficulties with voice control. Upper respiratory tract mucosal lesions may be present, as well as increased mucus production and coughing up blood. Both pulmonary and extrapulmonary involvement is common.

Up to 70% of cases have mucosal involvement, with lesions often found in the mouth, oropharynx, larynx, and palate. Classic lesions are superficial painful granular ulcers, with small spots of bleeding.

Cause

Paracoccidioidomycosis is caused by two species of fungi that can exist as a mold or yeast depending on temperature, *P. brasiliensis* and *P. lutzii*. In protected soil environments, near water sources, that are disturbed either naturally or by human activity, *P. brasiliensis* has been epidemiologically observed (although not isolated). A known animal carrier is the armadillo. In the natural environment, the fungi are found as filamentous structures, and they develop infectious spores known as conidia.

Human to human transmission has never been proven.

Mechanism

Primary infection, although poorly understood due to lack of data, is thought to occur through inhalation of the conidia through the respiratory tract, after inhaling fungal conidia produced by the mycelial form of *P. brasiliensis*. This occurs predominantly in childhood and young adulthood, after exposure to agricultural activity. Infection may occur through direct skin inoculation, although this is rare.

After inhalation into the alveoli, there is rapid multiplication of the organism in the lung tissue, sometimes spreading via the venous and lymphatic systems. Approximately 2% of people develop clinical features after the initial asymptomatic infection.

The type of immune response determines the clinical manifestation of the infection, with children and HIV co-infected individuals most commonly developing the acute/subacute disseminated disease. Most of those infected develop a Type 1 T-cell (T_h1) mediated immune response, resulting in alveolitis and compact granuloma formation that control fungal replication, and latent or asymptomatic infection. It then is thought to remain dormant in residual lung lesions and mediastinal lymph nodes. A deficient

T_h1 cell response results in the severe forms of the disease. In these individuals, granulomas do not form, and the affected person develops T_h2 and T_h9 responses, resulting in activation of B lymphocytes, high levels of circulating antibodies, eosinophilia, and hypergammaglobulinemia.

Lung involvement subsequently occurs after a dormant phase, manifesting in upper respiratory tract symptoms, and lung infiltrates on imaging. The commonest, chronic form, is almost certainly a reactivation of the disease, and may develop into progressive scarring of the lungs (pulmonary fibrosis).

It can cause disease in those with normal immune function, although immunosuppression increases the aggressiveness of the fungus. It rarely causes disease in fertile-age women, probably due to a protective effect of estradiol.

Diagnosis

A culture of Paracoccidioides brasiliensis during its yeast phase.

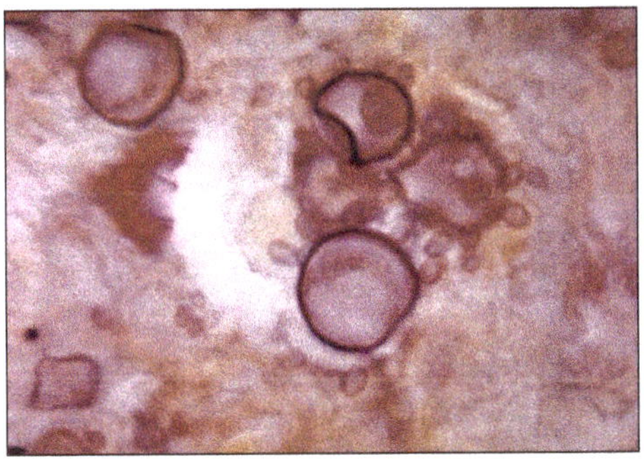

Histopathology of paracoccidiodomycoisis.

More than 90% of cases can be diagnoses with direct histological examination of tissue, such as sputum, bronchial lavage fluid, exudates and biopsies. Histopathological study with Gomori methenamine silver (GMS) stain or hematoxylin and eosin (H&E) stain revealing large yeast cells with translucent cell walls with multiple buds.

In the juvenile form, lung abnormalities are shown in high-resolution CT scans of the lungs, whereas in the chronic form plain X-rays may show interstitial and alveolar infiltrates in the central and lower lung fields.

Culture of *P. brasiliensis* takes between 20–30 days, requiring multiple samples and culture media. Initial culture can occur at room temperature, however after growth is noted, confirmation occurs by incubating at to 36-37 degrees to transform the fungus into yeast cells.

Antibody detection is useful both for acute diagnosis and monitoring. Gel immunodiffusion is commonly used in endemic areas, diagnoses 95% of cases with high specificity. Complement fixation allows for a measure of severity of cases by quantifying the antibody level, and is thus useful for monitoring treatment response. It is however only sensitive for 85% of cases, and cross-reacts with *H. capsulatum*.

Treatment

Both *P. brasiliensis* and *P. lutzii* are in-vitro susceptible to most antifungal agents, unlike other systemic fungal infections. Mild and moderate forms are treated with itraconazole for 9 to 18 months, as this has been shown to be more effective, has a shorter treatment duration and is more tolerated. Acidic beverages have been shown to reduce absorption of itraconazole. Co-trimoxazole is a second line agent, and is preferred for those with brain involvement, and during pregnancy.

For severe cases, intravenous treatment with amphotericin B is indicated, for an average of 2 to 4 weeks.

Prednisolone prescribed at the same time may reduce inflammation during treatment.

Patients should be treated until stabilisation of symptoms, and increase in body weight. Advice in regards to nutritional support, as well as smoking and alcohol intake should be provided. Treatment of adrenal insufficiency, if found, is recommended.

Clinical criteria for cure includes the absence or healing of lesions, stabilisation of body weight, negative as well as negative autoantibody tests.

ONYCHOMYCOSIS

Onychomycosis, also known as tinea unguium, is a fungal infection of the nail. Symptoms may include white or yellow nail discoloration, thickening of the nail, and separation of the nail from the nail bed. Toenails or fingernails may be affected, but it is more common for toenails to be affected. Complications may include cellulitis of the lower leg. A number of different types of fungus can cause onychomycosis including dermatophytes and *Fusarium*. Risk factors include athlete's foot, other nail diseases, exposure to someone with the condition, peripheral vascular disease, and poor immune function. The diagnosis is generally suspected based on the appearance and confirmed by laboratory testing.

Onychomycosis does not necessarily require treatment. The antifungal medication, terbinafine, taken by mouth appears to be the most effective but is associated with liver problems. Trimming the affected nails when on treatment also appears useful. There is a ciclopirox-containing nail polish, but it does not work as well. The condition returns in up to half of cases following treatment. Not using old shoes after treatment may decrease the risk of recurrence.

It occurs in about 10 percent of the adult population. Older people are more frequently affected. Males are affected more often than females. Onychomycosis represents about half of nail disease. It was first determined to be the result of a fungal infection in 1853 by Georg Meissner.

Signs and Symptoms

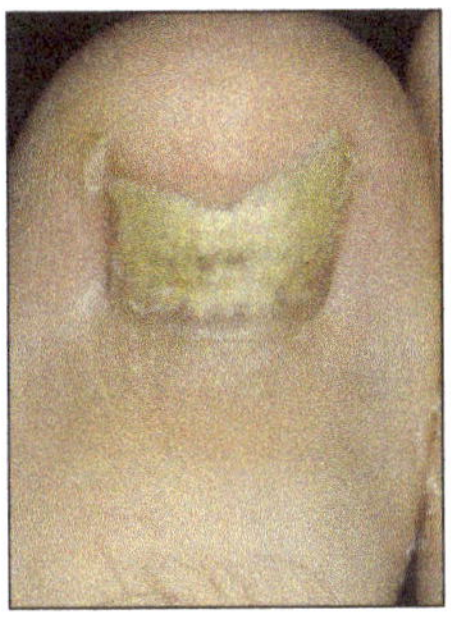

A case of fungal infection of the big toe.

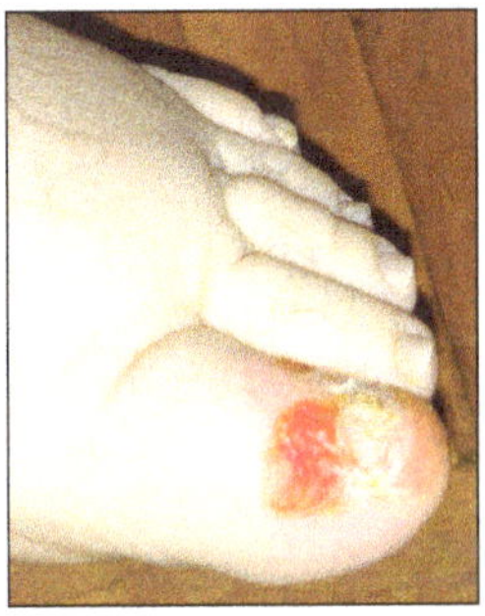

Advanced fungal infection of the big toe.

The most common symptom of a fungal nail infection is the nail becoming thickened and discoloured: white, black, yellow or green. As the infection progresses the nail can become brittle, with pieces breaking off or coming away from the toe or finger completely. If left untreated, the skin underneath and around the nail can become inflamed and painful. There may also be white or yellow patches on the nailbed or scaly skin next to the nail, and a foul smell. There is usually no pain or other bodily symptoms, unless the disease is severe. People with onychomycosis may experience significant psychosocial problems due to the appearance of the nail, particularly when fingers – which are always visible – rather than toenails are affected. Dermatophytids are fungus-free skin lesions that sometimes form as a result of a fungus infection in another part of the body. This could take the form of a rash or itch in an area of the body that is not infected with the fungus. Dermatophytids can be thought of as an allergic reaction to the fungus.

Causes

The causative pathogens of onychomycosis are all in the fungus kingdom and include dermatophytes, *Candida* (yeasts), and nondermatophytic molds. Dermatophytes are the fungi most commonly responsible for onychomycosis in the temperate western countries; while *Candida* and nondermatophytic molds are more frequently involved in the tropics and subtropics with a hot and humid climate.

Dermatophytes

When onychomycosis is due to a dermatophyte infection, it is termed tinea unguium. *Trichophyton rubrum* is the most common dermatophyte involved in onychomycosis. Other dermatophytes that may be involved are *T. interdigitale*, *Epidermophyton floccosum*, *T. violaceum*, *Microsporum gypseum*, *T. tonsurans*, and *T. soudanense*. A common outdated name that may still be reported by medical laboratories is *Trichophyton mentagrophytes* for *T. interdigitale*. The name *T. mentagrophytes* is now restricted to the agent of favus skin infection of the mouse; though this fungus may be transmitted from mice and their danders to humans, it generally infects skin and not nails.

Other Causative Pathogens

These include *Candida* and nondermatophytic molds, in particular members of the mold genus *Scytalidium* (name recently changed to *Neoscytalidium*), *Scopulariopsis*, and *Aspergillus*. *Candida* species mainly cause fingernail onychomycosis in people whose hands are often submerged in water. *Scytalidium* mainly affects people in the tropics, though it persists if they later move to areas of temperate climate.

Other molds more commonly affect people older than 60 years, and their presence in the nail reflects a slight weakening in the nail's ability to defend itself against fungal invasion.

Risk Factors

Aging is the most common risk factor for onychomycosis due to diminished blood circulation, longer exposure to fungi, and nails which grow more slowly and thicken, increasing susceptibility to infection. Nail fungus tends to affect men more often than women, and is associated with a family history of this infection.

Other risk factors include perspiring heavily, being in a humid or moist environment, psoriasis, wearing socks and shoes that hinder ventilation and do not absorb perspiration, going barefoot in damp public places such as swimming pools, gyms and shower rooms, having athlete's foot (tinea pedis), minor skin or nail injury, damaged nail, or other infection, and having diabetes, circulation problems, which may also lead to lower peripheral temperatures on hands and feet, or a weakened immune system.

Diagnosis

The diagnosis is generally suspected based on the appearance and confirmed by laboratory testing. The four main tests are a potassium hydroxide smear, culture, histology examination, and polymerase chain reaction. The sample examined is generally nail scrapings or clippings. These being from as far up the nail as possible.

Nail plate biopsy with periodic acid-Schiff stain appear more useful than culture or direct KOH examination. To reliably identify nondermatophyte molds, several samples may be necessary.

Classification

There are four classic types of onychomycosis:

- Distal subungual onychomycosis is the most common form of *tinea unguium* and is usually caused by *Trichophyton rubrum*, which invades the nail bed and the underside of the nail plate.
- White superficial onychomycosis (WSO) is caused by fungal invasion of the superficial layers of the nail plate to form "white islands" on the plate. It accounts for around 10 percent of onychomycosis cases. In some cases, WSO is a misdiagnosis of "keratins granulations" which are not a fungus, but a reaction to nail polish that can cause the nails to have a chalky white appearance. A laboratory test should be performed to confirm.
- Proximal subungual onychomycosis is fungal penetration of the newly formed nail plate through the proximal nail fold. It is the least common form of *tinea unguium* in healthy people, but is found more commonly when the patient is immunocompromised.
- Candidal onychomycosis is *Candida* species invasion of the fingernails, usually occurring in persons who frequently immerse their hands in water. This normally requires the prior damage of the nail by infection or trauma.

References

- Mycosis, science: britannica.com, Retrieved 16 July, 2020
- Mycoses: mycology.adelaide.edu.au, Retrieved 10 March, 2020
- FlorCruz, Nilo Vincent; Evans, Jennifer R (2015-04-09). Cochrane Eyes and Vision Group (ed.). "Medical interventions for fungal keratitis". Cochrane Database of Systematic Reviews. doi:10.1002/14651858.CD004241.pub4
- Barros MB, de Almeida Paes R, Schubach AO (October 2011). "Sporothrix schenckii and Sporotrichosis". Clinical Microbiology Reviews. 24 (4): 633–54. doi:10.1128/CMR.00007-11. PMC 3194828. PMID 21976602
- Castillo CG, Kauffman CA, Miceli MH (March 2016). "Blastomycosis". Infectious Disease Clinics of North America. 30 (1): 247–64. doi:10.1016/j.idc.2015.10.002. PMID 26739607
- Kauffman, CA (January 2007). "Histoplasmosis: a clinical and laboratory update". Clinical Microbiology Reviews. 20 (1): 115–132. doi:10.1128/CMR.00027-06. PMC 1797635. PMID 17223625
- Denning, D. W.; Pleuvry, A.; Cole, D. C. (March 2013). "Global burden of chronic pulmonary aspergillosis complicating sarcoidosis". European Respiratory Journal. 41 (3): 621–6. doi:10.1183/09031936.00226911. PMID 22743676
- Rigopoulos, Dimitris; Elewski, Boni; Richert, Bertrand (2018). Onychomycosis: Diagnosis and Effective Management. John Wiley & Sons. ISBN 9781119226505

PERMISSIONS

All chapters in this book are published with permission under the Creative Commons Attribution Share Alike License or equivalent. Every chapter published in this book has been scrutinized by our experts. Their significance has been extensively debated. The topics covered herein carry significant information for a comprehensive understanding. They may even be implemented as practical applications or may be referred to as a beginning point for further studies.

We would like to thank the editorial team for lending their expertise to make the book truly unique. They have played a crucial role in the development of this book. Without their invaluable contributions this book wouldn't have been possible. They have made vital efforts to compile up to date information on the varied aspects of this subject to make this book a valuable addition to the collection of many professionals and students.

This book was conceptualized with the vision of imparting up-to-date and integrated information in this field. To ensure the same, a matchless editorial board was set up. Every individual on the board went through rigorous rounds of assessment to prove their worth. After which they invested a large part of their time researching and compiling the most relevant data for our readers.

The editorial board has been involved in producing this book since its inception. They have spent rigorous hours researching and exploring the diverse topics which have resulted in the successful publishing of this book. They have passed on their knowledge of decades through this book. To expedite this challenging task, the publisher supported the team at every step. A small team of assistant editors was also appointed to further simplify the editing procedure and attain best results for the readers.

Apart from the editorial board, the designing team has also invested a significant amount of their time in understanding the subject and creating the most relevant covers. They scrutinized every image to scout for the most suitable representation of the subject and create an appropriate cover for the book.

The publishing team has been an ardent support to the editorial, designing and production team. Their endless efforts to recruit the best for this project, has resulted in the accomplishment of this book. They are a veteran in the field of academics and their pool of knowledge is as vast as their experience in printing. Their expertise and guidance has proved useful at every step. Their uncompromising quality standards have made this book an exceptional effort. Their encouragement from time to time has been an inspiration for everyone.

The publisher and the editorial board hope that this book will prove to be a valuable piece of knowledge for students, practitioners and scholars across the globe.

INDEX